POPULATION, REPRODUCTION AND
FERTILITY IN MELANESIA

Fertility, Reproduction and Sexuality

GENERAL EDITORS:

David Parkin, *Director of the Institute of Social and Cultural Anthropology, University of Oxford*
Soraya Tremayne, *Co-ordinating Director of the Fertility and Reproduction Studies Group and Research Associate at the Institute of Social and Cultural Anthropology, University of Oxford, and a Vice-President of the Royal Anthropological Institute*

Volume 1
Managing Reproductive Life: Cross-Cultural Themes in Fertility and Sexuality
Edited by Soraya Tremayne

Volume 2
Modern Babylon? Prostituting Children in Thailand
Heather Montgomery

Volume 3
Reproductive Agency, Medicine and the State: Cultural Transformations in Childbearing
Edited by Maya Unnithan-Kumar

Volume 4
A New Look at Thai AIDS: Perspectives from the Margin
Graham Fordham

Volume 5
Breast Feeding and Sexuality: Behaviour, Beliefs and Taboos among the Gogo Mothers in Tanzania
Mara Mabilia

Volume 6
Ageing without Children: European and Asian Perspectives on Elderly Access to Support Networks
Philip Kreager and Elisabeth Schröder-Butterfill

Volume 7
Nameless Relations: Anonymity, Melanesia and Reproductive Gift Exchange between British Ova Donors and Recipients
Monica Konrad

Volume 8
Population, Reproduction and Fertility in Melanesia
Edited by Stanley J. Ulijaszek

POPULATION, REPRODUCTION AND FERTILITY IN MELANESIA

Edited by
Stanley J. Ulijaszek

Berghahn Books
New York • Oxford

First published in 2005 by
Berghahn Books
www.BerghahnBooks.com

© 2005, 2008 Stanley J. Ulijaszek
First paperback edition published in 2008

All rights reserved. Except for the quotation of short passages for the purposes of criticism and review, no part of this book may be reproduced in any form or by any means, electronic or mechanical, including photocopying, recording, or any information storage and retrieval system now known or to be invented, without written permission of the publisher.

Library of Congress Cataloging-in-Publication Data

Population, reproduction, and fertility in Melanesia/edited by Stanley J. Ulijaszek.
     p. cm. -- (Fertility, reproduction, and sexuality ; v. 8)
   Includes index.
   ISBN 978-1-57181-644-3 (hbk), 978-1-84545-269-8 (pbk)
   1. Demographic anthropology--Melanesia. 2. Fertility, Human--Melanesia. 3. Human reproduction--Melanesia. 4. Melanesia--Population. 5. Melanesia--Social life and customs. I. Ulijaszek, Stanley J. II. Series.

GN668.P67 2005
304.6'0995--dc22
                                                  2005043634

British Library Cataloguing in Publication Data

A catalogue record for this book is available from the British Library.

Printed in the United States on acid-free paper

ISBN 978-1-57181-644-3 (hbk), 978-1-84545-269-8 (pbk)

Contents

List of Figures and Tables   vii

List of Contributors   ix

Introduction: Population Change, Social Reproduction and Local Understandings of Fertility in Melanesia   1
Stanley J. Ulijaszek

1. Fertility and the Depopulation of Melanesia: Childlessness, Abortion and Introduced Disease in Simbo and Ontong Java, Solomon Islands   13
Tim Bayliss-Smith

2. The Impacts of Colonialism on Health and Fertility: Western New Britain 1884–1940   53
C. Gosden

3. Purari Population Decline and Resurgence across the Twentieth Century   67
Stanley J. Ulijaszek

4. Migration and Fertility of a Small Island Population in Manus: a Long-term Analysis of its Sedentes and Migrants   90
Yuji Ataka and Ryutaro Ohtsuka

5. Fertility and Social Reproduction in the Strickland-Bosavi Region   110
Monica Minnegal and Peter D. Dwyer

6. 'Emptiness' and Complementarity in Suau Reproductive Strategies   136
Melissa Demian

7. Cognitive Aspects of Fertility and Reproduction in Lak, New Ireland 159
   Sean Kingston

8. History Embodied: Authenticating the Past in the New Guinea Highlands 182
   Michael O'Hanlon

9. Variations on a Theme: Fertility, Sexuality and Masculinity in Highland New Guinea 201
   Pascale Bonnemère

10. Fertility among the Anga of Papua New Guinea: a Conspicuous Absence 218
    Pierre Lemonnier

Index 239

List of Figures and Tables

Figures

1.1.	Island Melanesia excluding the Fiji group, showing the names of islands mentioned in the text	16
2.1.	Language groups on the southern coast of West New Britain	57
3.1.	Infant mortality rates among the Wopkaimin and Mount Obree populations, 1982–1988	74
3.2.	Population pyramid, Kinipo residents, 1955	77
3.3.	Population pyramid, Kinipo residents, 1996	78
3.4.	Age-specific fertility rates in Papua New Guinea	82
3.5.	Age-specific fertility and number of years of education, Purari delta	83
3.6.	Fertility and income, Purari delta	84
4.1.	Location of the Baluan Island and Manus Province, Papua New Guinea	92
4.2.	Perelik population size, 1955–1995	97
4.3.	Change in the number of persons in the urban sector from 1955 to 1995, according to occupational status	99
4.4.	Age-specific fertility rate (ASFR) of women in rural urban sectors in the four 10-year periods	101
5.1.	Map of study area showing language groups and localities	112
5.2.	Kinship connections among residents of (A) the Kubo village of Gwaimasi and (B) the Bedamuni village of Ga:misi at January 1999	114
6.1.	Exchange between *iha*	145
6.2.	'Consolidating' an adoption	152
6.3.	'Dissolving' an adoption	152
7.1.	This mask is a *nantoi*, a mother tubuan, the apical image of the men's initiatory cults and the most fearsome yet seductive sight in Lak	165

7.2.	Dancers with *kabut*, lesser versions of tubuan	166
7.3.	Exchanges at *tondong*	169
7.4.	*Nantoi* tubuan remove 'bones' of the deceased	172
7.5.	A *dal* in her spiritual decoration	174
7.6.	A less formed male spirit (*tamsaikio*) appears at night	176
7.7.	*Malerra*, men with various owned spiritual decorations and qualities	177
8.1.	Yimbal Aipe (later member of parliament for North Wahgi) dancing in the Pig Festival, 1979	186
8.2.	Moru and daughter, 1980	197

Tables

1.1.	The depopulation of Island Melanesia: the conventional model	25
1.2.	Fertility on Simbo for three generations	29
1.3.	The reproductive histories of 110 Ontong Java women aged over 30 years in 1972	43
1.4.	Ontong Java cohorts for perceived pregnancies and for live births, showing mortality up to age of 20, by time period, 1921–1972	45
4.1.	Infant mortality rate, life expectancy at birth, and annual increase rate, broken down by the four regions of Papua New Guinea	93
4.2.	Annual population increase rate in the rural and urban sectors for four 10-year periods and the whole duration, 1955–1994	97
4.3.	Number of gardens in use by all households of the 10 Perelik lineages, broken down by own land and borrowed land, and number of gardens in use by other households in the 10 lineages' land	105
5.1.	Fertility of Bedamuni and Kubo-Konai women	115
6.1.	Samalua and Elelia's adoptees	141
6.2.	Joshua and Paini's adoptees	142

List of Contributors

Dr Y. Ataka, School of Policy Studies, Kwansei Gakuin University, 2-1 Gakuen, Sanda 669-1337, Japan.

Dr T. Bayliss-Smith, Reader, Department of Geography, University of Cambridge, Downing Place, Cambridge CB2 3EN, United Kingdom.

Dr P. Bonnemère, CNRS Centre de Recherche et de Documentation sur l'Oceanie, Universite de Provence, Campus Saint-Charles, 3 Place Victor-Hugo, 13003 Marseille, France.

Dr M. Demian, Department of Anthropology, Emory University, 1557 Dickey Drive, Atlanta GA 30322, USA.

Dr P.D. Dwyer, School of Anthropology, Geography and Environmental Studies, University of Melbourne, Victoria 3010, Australia.

Professor C. Gosden, Pitt-Rivers Museum, University of Oxford, South Parks Road, Oxford, United Kingdom.

Dr S. Kingston, Independent Scholar and Publisher, 57 Orchard Way, Wantage, Oxfordshire OX12 8ED, United Kingdom.

Professor P. Lemonnier, CNRS Centre de Recherche et de Documentation sur l'Oceanie, Universite de Provence, Campus Saint-Charles, 3 Place Victor-Hugo, 13003 Marseille, France.

Dr M. Minnegal, School of Anthropology, Geography and Environmental Studies, University of Melbourne, Victoria 3010, Australia.

Professor R. Ohtsuka, President, National Institute for Environmental Studies, 16-2 Onogawa, Tsukuba, 305-8506, Japan.

Dr M. O'Hanlon, Pitt-Rivers Museum, University of Oxford, South Parks Road, Oxford, United Kingdom.

Professor S. Ulijaszek, Institute of Social and Cultural Anthropology, University of Oxford, 51 Banbury Road, Oxford OX2 6PF, United Kingdom.

INTRODUCTION

POPULATION CHANGE, SOCIAL REPRODUCTION AND LOCAL UNDERSTANDINGS OF FERTILITY IN MELANESIA

Stanley J. Ulijaszek

Three intersecting themes – population, fertility and reproduction – form the basis of this volume. What prompted this collection of essays was the relative lack of recent literature concerning these issues in Melanesia. The majority of the articles published here were papers given in the Fertility and Reproduction Study Group Seminar Series 'Fertility and Reproduction in Melanesia', at the Institute of Social and Cultural Anthropology, University of Oxford. The title of the seminar series acknowledged the historical contribution of W.H.R. Rivers (1922a) to population studies in this region, and to update what is known of population size and process since that time. Given that much has changed in the 82 years since Rivers' publication, a primary focus for both the seminar series and the volume, was, and is, historical demography: the decline and subsequent population resurgence in Melanesia.

While population processes reveal the relative biological success or failure of a society, an examination of ideas of reproduction from the perspectives of local communities leads to quite different understandings of this term. The question of exactly what is being reproduced, and how, is the second theme of the volume: how is the social reproduced, and how is the cultural uncoupled from the biological in various societies in this region. The third issue, that of local understandings of fertility, is an evaluation by various authors of the now-extensive literature concerning human life cycles and how they are structured culturally. Thus the volume is at one and the same time an evaluation of past population processes, a 'bringing up-to-date' of population processes across the twentieth century, and a survey of local understandings of fertility and reproduction in broader social and biological contexts.

Thus the volume goes beyond Rivers' concerns about population decline in Melanesia, to examine population, fertility and reproduction in the New Guinea Highlands, a region uncontacted by Europeans at the time of Rivers' volume. It also examines the relationship between cultural and biological processes that structure the reproduction of populations and societies, and it is able to examine local understandings in a way that was not possible in the early twentieth century. Ideas of what the region constitutes, and why, have also been changing, making the naming of the region for the purposes of the volume problematic. There are various ways of constructing the broader region considered here: 'Melanesia', 'Near Oceania', 'Austronesia', 'New Guinea and the Solomon Islands' among them.

I would like to discuss the idea of 'Melanesia' as a region, before going on to describe the contributions of the various authors to this volume. Tim Bayliss-Smith (chapter 1) identifies the key features of 'Melanesia' as a geographical construct of nineteenth-century geography, which came after European acknowledgement in the late eighteenth century that 'Polynesia' was a region with a common language and therefore, also, with a common culture. In 1833 the navigator Dumont d'Urville distinguished between Polynesia (the many islands), Melanesia (the black islands), and Micronesia (the small islands). Melanesians were characterised by their blackness, linguistic diversity, and their little-developed political institutions. D'Urville's Melanesia originally included Australia, as well as New Guinea, Fiji, Vanuatu and New Caledonia. As Australia became increasingly settled by the British, it was reclassified as a separate entity to the other black islands. Thus the colonial process, and nineteenth-century understandings of population and geography underpinned the naming of the region 'Melanesia'. In this volume, I fall to a pragmatic retention of the term 'Melanesia' as the overarching regional descriptor, notwithstanding the problems associated with such regionally-delimiting nomenclature.[1]

The notion of Melanesia as a geographical region is preserved here in an historical sense that allows the changing demography of the societies within this region, as discussed in Rivers (1922a), and by Roberts (1927), Hogbin (1930, 1939) and others, to be considered within the framework of early colonial encounters. There is no intent to use the term in any social evolutionist sense at any place in the pages and chapters that follow. In this volume, various authors use a number of overlapping regional classifications, according to their analytic utility as historical, geographical, political and culture-area delimiters: 'Melanesia', 'Polynesia', 'Austronesia', 'New Guinea', 'Papua New Guinea', 'Islands Melanesia', 'Islands region', 'Highlands region', 'New Guinea Highlands', and 'South Coast New Guinea' among them. The term 'Melanesia' is used as an umbrella which I hope allows area comparisons of various kinds – historical, colonial, regional, cultural – to be made by authors where useful. That said, some authors do not rely on such analytical devices, favouring instead to speak to specialised interests central to, or involving demography, fertility and reproduction, but which address more general anthropological problems (Brady 1989).

The first theme of this book is addressed in the first three chapters of this volume, while the issue of subsequent population increase is also described and discussed in the same first three chapters and the one that follows them. The demise of the people of geographical Melanesia was summarised very succinctly by Durrad (1922) writing in W.H.R. Rivers' (1922a) edited volume *Essays on the Depopulation of Melanesia*:

> anyone who has spent a few years in Melanesia will have noticed, between the time of his arrival and that of his departure, a distinct difference in the number of people among whom he has lived. The longer his stay extends, the more marked becomes the fall in population.

Various explanations are given by Rivers (1922b) for the population demise reported for many parts of this geographical region. These include death by way of epidemics, low birth rate because of physical dislocation caused by the European plantation labour trade, sexually transmitted diseases (STDs) and the so-called 'psychological factor' (the loss of morale and of the will to live among local people after European colonization) that Rivers (1922b) believed to be the principal reason for population decline. Fertility decline was viewed by him to have been of lower importance in population decline than increased mortality, with the introduction of infectious disease. However, the introduction of one category of infectious disease, STDs, was seen as detrimental to fertility.

Durrad (1922) and Roberts (1927) believed that Melanesian populations were already in decline with the earliest European traders to New Guinea, and what they were observing in the early twentieth century was but an acceleration of the process. While we may never be certain of this, we can at least update the population history of various communities in this region and re-examine the classic accounts of population decline and its' causes, as given by Rivers (1922b), Roberts (1927) and Hogbin (1930, 1939).

In the first of the three chapters in which the historical demography of the region are considered, Bayliss-Smith contextualizes what is to follow by giving a comprehensive account of population processes in the Pacific region generally, after the 'fatal impacts' of colonial encounters (Moorehead 1966) in the nineteenth and early twentieth centuries. He goes on to reconstruct demographic processes in the Solomon Islands to examine Rivers' (1922b) explanations for the processes of depopulation there. Using Rivers' data set from Simbo, western Solomon Islands, he constructs a chronology for demographic change in this region, across the period 1790 to 1910. The absence of clear evidence of epidemics of introduced disease, and of other factors commonly cited for the depopulation of Melanesia, had lead Rivers to conclude that 'the psychological factor' was central to high mortality and fertility decline there. Bayliss-Smith is able to adduce enough new information in this chapter to suggest that Rivers was incorrect, by strongly implicating the European introduction of infectious diseases, both epidemic and endemic (in the case of STDs), as the prime factor in the population decline on Simbo.

According to Bayliss-Smith, depopulation of Simbo probably began before 1850, and was not reversed until the mid-twentieth century, largely because of health interventions which have reduced mortality rates, a trend observed elsewhere in the region, including the Purari, in South Coast New Guinea (SCNG) (Ulijaszek, Chapter 4). Bayliss-Smith contrasts the Simbo case with that of depopulation on the Polynesian atoll of Ontong Java, located 270 km north of Santa Isabel in the Solomon Islands, where population reached its lowest size in 1939. Earlier, Hogbin (1930) attributed population decline on Ontong Java to an enhanced death rate, due to a combination of psychosomatic and pathological causes, the impacts of infectious disease being heightened by the people's 'state of mind which acquiesces in extinction' (Hogbin 1930).

Bayliss-Smith reinterprets the Ontong Java demographic data to suggest that the introduction of infectious disease, malaria in particular, is largely adequate to explain the population decline. As with Simbo, the decline in the Ontong Java population persisted until the middle of the twentieth century. Reasons given by Bayliss-Smith for the subsequent population recovery include the relative isolation of the atoll (which had been a trading nexus) during the Second World War, the decline in epidemics of infections in the post-war period, and malaria control in the 1960s. In 1986 the resident population began to approach its nineteenth-century size, with about 20 percent of the population living in urban conditions either in the capital of the Solomon Islands, Honiara, or elsewhere (Bayliss-Smith 1986). This pattern of high urbanism in association with population resurgence is also demonstrated in this volume for the Purari (Ulijaszek, chapter 4) and Perelik villagers of the Manus Islands (Ataka and Ohtsuka, chapter 5).

In the second of the historical demography articles, Chris Gosden considers the idea of the 'fatal impact' of European colonisation (Moorehead 1966), implicit in the writings of Rivers (1922b), Roberts (1927) and Hogbin (1939), in respect of colonialism, health and fertility change in western New Britain between 1884 and 1940. The population of this region declined dramatically in the 1880s and 1890s, soon after European settlement there. While confirming the fatal impact view of population decline, Gosden argues that the biological threat of new diseases (including high mortality and depopulation) elicited responses of unprecedented nature among local communities, leading him to reject the view that they largely remained passive in relation to colonisation. Among the responses to high mortality and demographic change was physical reorganisation, with the formation of new villages, and the introduction and use of new crops and technologies. To some extent, similar types of response to European colonisation also took place among the Purari (Maher 1961) (Ulijaszek, chapter 4).

The population shocks associated with colonialism are also considered by Stanley Ulijaszek, in his chapter on population decline and subsequent resurgence among the Purari of SCNG. He argues that the dramatic reduction in population size in the early twentieth century can be attributed largely to the introduction of infectious diseases by Europeans, and to low crude birth rate associated with the recruitment of adult males for plantation labour. As with the Simbo and Ontong Java populations (Bayliss-Smith, chapter 1), demographic decline persisted into the middle of the twentieth century. Subsequent population increase is likely to have been due to the introduction and increased availability of biomedicine, leading to reduced infectious disease, as well as to improved nutrition. Ulijaszek argues that improved nutrition has lead to greater fecundity of women, while both improved nutrition and reduced infectious disease have lead to improved survivorship of young children. Among the Purari, high total fertility rates (TFR) have been maintained post-independence, despite economic change and modernization. Analysis of data collected by Ulijaszek in 1995 and 1997 indicate that greater cash income is positively associated with both TFR and earlier age at which women start to bear children, indicating a tendency for economically successful males to take younger brides. Thus, material well-being translates into a larger number of offspring, rather than increased investment in a smaller number of offspring. Although this runs counter to conventional demographic transition theory, this pattern is found elsewhere in New Guinea, among a Mountain Ok group undergoing rapid economic change (Taufa et al. 1990). With population increase has come increased migration of the Purari population to urban centres, especially Port Moresby. However, although the urban Purari population is likely to have exceeded 30 percent of the total Purari population in 1996, urban connectedness, through urban relatives, or urban migration at some time of life, seems to have no influence on TFR, despite the expectation that exposure to outside ideas concerning appropriate family size might be a force for reducing fertility rates.

In the fourth of the historical demography chapters, Ataka and Ohtsuka consider population processes in a local population in the Manus Islands, in and from Perelik village, Baluan Island. They descibe population growth, rural to urban migration, and fertility rates of both rural and urban groups, between 1955 and 1995. In general, the Manus population has been more acculturated, and from an earlier time, than elsewhere in Island New Guinea. As in west New Britain (Gosden, chapter 2), Ontong Java (Bayliss-Smith, chapter 1), and among the Purari (Ulijaszek, chapter 3), Komblo of the New Guinea Highlands (O'Hanlon, chapter 8), and Simbo (Bayliss-Smith, chapter 1), Perelik population growth has been considerable in the second half of the twentieth century. This is attributed by Ataka and Ohtsuka to the control of infectious disease, especially malaria, in combination with high TFR, at least until the mid-1970s. As with Ontong Java in the early twentieth century (Bayliss-Smith, chapter 1), the prevalence of STDs on Manus was low in the years following the

Second World War, and does not appear to have been a factor influencing fertility.

The growth of the urban Perelik population has been exceptional; toward the end of the twentieth century, over 40 percent of all Perelik people lived in urban centres, a much higher value than for either the Ontong Java (about 20 percent) or Purari (just over 30 percent) populations. The increased ratio of urban to rural Perelik after about 1985 is attributed by Ataka and Ohtsuka to the maintenance of a high TFR in the urban population, at a time when the TFR in Perelik village showed considerable decline, against a background of strong infectious disease control in both locations. This runs counter to expectations, but is explained by Ataka and Ohtsuka in the following way. Population pressure associated with a shortage of land in Perelik village might have contributed to greater uptake of family planning in the rural sector, as well as contributing to rural to urban migration. There may have been less concern to limit family size in the urban sector, where Perelik male heads of household were mostly public servants, teachers and office workers, whose earnings may have been adequate to rear many children.

Ataka and Ohtsuka also note the emergence of social and economic problems associated with Perelik population processes. In recent years, new job opportunities in urban areas have been few and the relative educational advantage of Perelik people in finding work is diminishing. The number of urban Perelik migrants without purpose has increased in recent years, creating a pool of urban dwellers without jobs. According to Ataka and Ohtsuka, the capability of the wantok system to absorb such migrants is limited, leading the authors to anticipate significant back-migration to Perelik from urban centres in coming years. This would increase the population density of Perelik village and exacerbate land shortage there. This is an issue that many populations in Melanesia may face in the future.

The next three chapters consider the second theme of the book, which is the different ways in which biological and social reproduction are uncoupled in Melanesian societies. In the first of these, Monica Minnegal and Peter Dwyer compare ideas of reproduction among two closely related Papua New Guinean societies, Kubo and Bedamuni, who live in neighbouring areas of the Strickland-Bosavi region. Dwyer (1996) proposed the idea that with increased intensification and decreased egalitarianism, understandings of 'culture' and 'nature' become progressively decoupled. In their chapter, Minnegal and Dwyer confirm this idea in relation to reproduction, by demonstrating that the more densely populated and socially complex Bedamuni prioritize social reproduction and pattern their biological outcomes to satisfy social desires to a greater extent, than do the more egalitarian Kubo. The authors conclude by urging caution concerning evolutionary ecological and sociobiological explanations of demographic processes, suggesting that they can only be partial if they do not accommodate the understanding that people often view social and biological reproduction as being disconnected arenas, living in ways such that the latter is incidental to the former.

The second of this cluster of three chapters, by Melissa Demian, is concerned with adoption among Suau people of Milne Bay Province, Papua New Guinea. While adoption can be understood in terms of natural relationships that are co-opted by cultural imperatives, Demian points out that the question of exactly what is reproduced by an adoption strategy is usually unasked by anthropologists. She shows that Suau adoption is an outcome of the desire to be part of social relationships, both past and present. A sibling set in Suau which is characterised by an absence of either boys or girls, is considered 'empty' by them because it does not permit the anticipation of adult relationships which depend on both same-sex and cross-sex axes of support and nurture. For Suau, it is not enough to reproduce persons: the right sorts of persons must be reproduced, so that the right sorts of relationships are reproduced. As gifts between adults in the removal of 'emptiness', Suau adoptees are important objectifications of relationships which are reproduced over time.

In the next chapter, Sean Kingston shows how death and birth are linked spiritually for Lak people in south-east New Ireland. Their understanding of the human life cycle, involving conception at birth and deconception at death has much in common with many Austronesian societies (Mosko 1983, 1989). Kingston argues that conception and birth are the reverse of 'de-conception', the disarticulation after death of the social relations that the person embodied. In Lak, birth and its rites are construed cognitively, spoken of as a 'remembering and bringing into mind' that contrasts with the 'forgetting and absenting from mind' of death. The 'conception' of babies at their birth only takes place via a disarticulation of the totalizing figure of the most powerful spirits that take away the final remains of the dead person, mirroring the forgetting of the dead as a coherent nexus of social relations and attention. Kingston goes on to argue that although the child is a rearticulation of someone whose social form has previously been disarticulated after death, they can only become a more determinate form through processes of attention. As children become adults and proceed through their life-course they continue to form themselves and others through the action of reciprocal attention. Without thinking of each other, people would have no form, and would therefore be unknown.

The final three chapters are concerned with local understandings of fertility and reproduction in New Guinea. In the first of these, Michael O'Hanlon shows how the Wahgi of the Western Highlands Province of Papua New Guinea view patterns of fertility and reproduction as part of a much broader universe of signs they turn to, to authenticate moral probity. Rival narratives, which purport to explain misfortunes as an enchained series whose origins lie back in relations with clanspeople, and with 'source' people like maternal kin, are manifold. Wahgi cosmology drives them to make sense of individual births and deaths as an interconnected series, reflecting hidden intra-clan treachery or breaches of proper relations across time.

In chapter 9, Pascale Bonnemère considers the diversity of themes concerning fertility, sexuality and masculinity in the New Guinea Highlands. She focuses on two major sets of rituals: bachelor cults and male initiations, showing that although they appear to be fairly contrasted in their organization and content, they can be viewed as variations on a single universal theme concerning female reproductive capacities. By comparing Ankave Anga male initiations and the Enga bachelor cult, Bonnemère shows that spiritual marriage does not seem to be limited to the bachelor cults found in the Highlands but appears to be a more general symbolic reference that may also occur in highland fringe societies, such as the Anga. Bonnemère's analysis points to an opposition between male initiations in which the symbolism of human growth is largely predominant and in which the novices' mothers (present in person or as surrogates) are essential figures of the ritual process, and spirit cults, in which fertility is ensured by the symbolic re-enactment of a coitus scene and in which the principal figure is a spiritual being, a virtual wife. Bonnemère argues that whether or not ritual forms in New Guinea aim to assure general fertility, the model referred to by ritual symbolism is most often that of human reproduction. According to her, the differences encountered across New Guinea societies are merely variations on this theme.

In the final chapter in which local understandings of fertility are considered, Pierre Lemonnier presents an interesting anomaly. Whereas many New Guinea societies are known for their fertility rituals, or for the place occupied by fertility in some of their outstanding institutions and therefore in people's everyday life, the Anga of Eastern Highlands, Gulf and Morobe provinces of PNG lack, and have lacked, any such collective practice. Lemonnier shows that they have limited their interest in the circulation of a life-force to very specific domains: the making of adult men and warriors during male initiations and, for some of them, the unspoken recycling of life-giving substance within clans or lineages. As for the fertility of women, the theme is strikingly absent among Anga. It is possible that this situation may not be unique in New Guinea, but Lemonnier shows that it contrasts strongly with the situation among both neighbouring Highlander groups and SCNG societies.

This volume spans the physical realities of population decline and subsequent resurgence, the ways in which the biological is uncoupled from the social in the context of reproduction, and some local understandings of reproduction in Melanesia. It cannot, and indeed does not, aim to be comprehensive in its treatment of any one of these areas. However, in undertaking such a broad-ranging survey, authors in this volume offer a range of biological and cultural generalities, and show some quite specific ways in which the cultural (Thomas 1989) and biological (Alpers and Attenborough 1992) diversities that Melanesia is known for, are clearly evident in population, reproduction and fertility.

Acknowledgements

I thank Tim Bayliss-Smith for discussion of the various terminologies possible as an umbrella term for the region, and to Peter Dwyer and Monica Minnegal for comments on an ealier draft of this manuscript.

Note

1. The idea of 'Melanesia' as a culture area has been contrasted with that of 'Polynesia' on the basis of local social and political organisation (Sahlins, 1963). However, this has been criticised on the basis of linguistic, archaeological and anthropological heterogeneity (Hau'ofa 1975; Pawley 1981; Guiart 1981; Thomas 1989). The issue of regionally-delimiting nomenclature has bothered me as editor, particularly in respect of the title to the volume, which has to accommodate the variety of classifications in current usage. The term 'Near Oceania', as the area incorporating New Guinea and the Solomon Islands, as a geographical region of great ethnographic diversity, but not associated with the nineteenth-century views of

race and language which were used in the creation of the idea of Melanesia, has been put forward by Green (1989). However, ambiguity remains, because the definition of Near Oceania is dependent on a provisional and extremely incomplete map of archaeological finds, while placing some populations, including those of the Santa Cruz Islands and the present-day nations of Vanuatu and New Caledonia, in 'Far Oceania' when they have greater similarity to populations considered to be part of Near Oceania (Bayliss-Smith, personal communication). In the case of the Santa Cruz Islands, the presence of Papuan languages suggests inclusion in Near Oceania, but the absence of Pleistocene settlement suggests inclusion in Far Oceania (Bayliss-Smith, personal communication).

References

Alpers, M. and R. Attenborough (eds). 1992. *The Small Cosmos. Human Biology of Papua New Guinea*. Oxford: Oxford University Press.

Bayliss-Smith, T.P. 1986. *Ontong Java Atoll: Population, Economy and Society, 1970–1986*, Occasional Paper 9, South Pacific Smallholder Project, University of New England: Armidale.

Brady, I. 1989. 'Comments on Paper by N. Thomas: the Force of Ethnology. Origins and Significance of the Melanesia/Polynesia division', *Current Anthropology* 30: 34–35.

Durrad, W.J. 1922. 'The Depopulation of Melanesia'. In *Essays on the Depopulation of Melanesia*, ed. W.H.R. Rivers, 3–24. Cambridge: Cambridge University Press.

Dwyer, P.D. 1996. 'The Invention of Nature'. In *Redefining Nature: Ecology, Culture and Domestication*, eds R. Ellen and K. Fukui, 157–186. Oxford: Berg.

Green, R.C. 1989. 'Comments on Paper by N. Thomas: the Force of Ethnology. Origins and Significance of the Melanesia/Polynesia Division', *Current Anthropology* 30: 35–36.

Guiart, J. 1981. 'A Polynesian Myth and the Invention of Melanesia', *Journal of the Polynesian Society* 90: 139–144.

Hau'ofa, E. 1975. 'Anthropology and the Pacific Islanders', *Oceania* 45: 283–289.

Hogbin, H.I.P. 1930. 'The Problem of Depopulation in Melanesia as Applied to Ongtong Java (Solomon Islands)', *Journal of the Polynesian Society* 39: 43–66.

———. 1939. *Experiments in Civilization: the Effects of European Culture on a Native Community of the Solomon Islands*. London: George Routledge.

Maher, R.F. 1961. *New Men of Papua*. Madison: University of Wisconsin Press.

Moorehead, A. 1966. *The Fatal Impact: an Account of the Invasion of the South Pacific, 1767–1840*. London: Hamilton.

Mosko, M. 1983. 'Conception, De-conception and Social Structure in Bush Mekeo Culture', *Mankind* 14: 23–32.

———. 1989. 'The Developmental Cycle among Public Groups', *Man* 24: 470–484.

Pawley, A. 1981. Melanesian Diversity and Polynesian Homogeniety. A Unified Explanation for Language. In *Studies in Pacific Languages and Cultures*, eds J. Hollyman and A. Pawley. Auckland: Linguistic Society of New Zealand.

Rivers, W.H.R., ed. 1922a. *Essays on the Depopulation of Melanesia*. Cambridge: Cambridge University Press.

———. 1922b. 'The Psychological Factor'. In *Essays on the Depopulation of Melanesia*, ed. W.H.R. Rivers, 84–113. Cambridge: Cambridge University Press.

Roberts, S.H. 1927. *Population Problems in the Pacific*. London: Routledge.

Sahlins, M. 1963. 'Poor Man, Rich Man, Big-man, Chief: Political Types in Melanesia and Polynesia', *Comparative Studies in Society and History* 5: 285–303.

Taufa, T., V. Mea and J. Lourie, 1990. 'A Preliminary Report on Fertility and Socio-economic Changes in Two Papua New Guinea Communities'. In *Fertility and Resources*, eds J. Landers and V. Reynolds, 35–46. Cambridge: Cambridge University Press.

Thomas, N. 1989. 'The Force of Ethnology. Origins and Significance of the Melanesia/Polynesia Division', *Current Anthropology* 30: 27–34.

CHAPTER 1

FERTILITY AND THE DEPOPULATION OF MELANESIA: CHILDLESSNESS, ABORTION AND INTRODUCED DISEASE IN SIMBO AND ONTONG JAVA, SOLOMON ISLANDS

Tim Bayliss-Smith

Depopulation and colonialism: the big picture

Melanesia was one of the last regions of the world to be affected by the process of global integration that effectively began in 1492 with the European colonisation of the New World. It was a process accomplished with the aid of 'Guns, Germs and Steel', to quote the title of Jared Diamond's (1997) account. The process spread to Australia and Polynesia in the late eighteenth century, reaching Fiji in about 1810 and some of the islands in western Melanesia in the 1840s. The 'scramble for the Pacific' by European colonial powers reached its climax in the 1890s, and apart from some small isolated pockets, global integration was completed in the 1930s with the Australian and Dutch expansion into the New Guinea highlands.

In all regions previously isolated from Eurasia and Africa, namely the Americas, Australasia and Oceania, the outcome of contact and colonisation was decided not only by 'guns and steel' but also by 'germs' – by disease pathogens and their demographic impacts (Diamond 1997). Before 1492 the indigenous peoples of this global periphery had an advantage over Eurasians and Africans in suffering from a smaller number of infectious diseases, but sustained contact transformed their situation. In the Americas, for example, Crosby (1993) lists the following new diseases introduced after 1492: smallpox, measles, chicken pox, whooping cough, typhus, typhoid fever, bubonic plague, cholera, scarlet fever, malaria, yellow fever, diphtheria and influenza. Whether or not a name should be added or subtracted from this list is relatively unimportant, Crosby suggests, in view of the 'avalanche of disease that decimated all native American peoples, and even obliterated many [societies], ... struck by the micro-invaders and the macro-invaders simultaneously' (Crosby 1993: 86).

The scale of depopulation in the Americas, and the process itself, have been much studied by historians in recent decades. The highland populations in the former Aztec and Inca empires are now estimated to have declined by at least 90 percent in the first post-contact century. Mexico's population, for example, fell from about 30 million in 1492 to less than 3 million in 1600 (Blaut 1993: 194). The decline in the Amerindian populations of the tropical lowlands and Caribbean islands is not so fully recorded, but often it was even more extreme.

In Polynesia an equivalent catastrophic decline has been proposed by Stannard (1989) for the population of Hawai'i in the century after 1778. McArthur (1968) studied the demographic histories of Tonga, Samoa, Cook Islands and French Polynesia, and came to more cautious conclusions. Her research suggested that only three of the Polynesian island groups she studied, Tahiti, Marquesas and Rarotonga, suffered really severe post-contact effects. Rarotonga's population was halved, mainly by dysentery, while Tahiti declined from about 35,000 in 1769 to under 8,000 one hundred years later. The Marquesas islands were probably even more severely hit, but the pre-contact population estimates may not be reliable (McArthur 1968: 281). Pirie (1972: 202) blamed gonorrhoea, spread by the crews of visiting whaling ships, for low fertility in the Marquesas in the nineteenth century.

The Australian evidence has also been critically re-examined (Reynolds 2001), and it shows that while a decline in the Aboriginal population was almost universal, both chronology and causation were quite variable. In the state of Victoria, for example, signs of smallpox had been noted in 1830 among the Aboriginal population even before European settlement was successfully established. Aborigines in Victoria numbered about 12,000–15,000 in 1830. Fewer than 2,000 survived to be counted in the first census, 33 years later (Flannery 1995). In the Northern Territory populations declined most rapidly from 1900 until 1940, but even in the twentieth century, accurate information is difficult to establish. Introduced infections took a heavy toll, especially sexually transmitted diseases (STDs) that also reduced birth rates. Humanitarians suspected that conditions of hygiene and nutrition were poor for Aborigines on the cattle stations, but the European pastoralists blamed infanticide for the depopulation (McGrath 1987).

A persistent feature of Australian Aboriginal demography at the time of its most rapid decline was the preponderance of males in the population. Children also made up a much smaller proportion of the population than is usual, even among European populations. The large number of men has never been satisfactorily explained, but the small number of children can be attributed to low fertility because of the adult sex imbalance, and to 'horrific infant mortality', respiratory diseases being the most common cause of death (Flannery 1995: 326). In the Northern Territory many trained observers commented on a state of apathy and depression, and noted the detrimental effects of the Aborigines' fatalistic view of illness. Many sick people and pregnant women declined to make use of the limited medical and natal facilities available. Appallingly high maternal and infant death rates were reported by the anthropologists R.M. and C.H. Berndt in the 1940s, and fertility rates were also low. At none of the cattle stations where the Berndts carried out surveys did the average number of children per woman exceed 2.3, while infant mortality rates ranged from 32 to 52 percent (Berndt and Berndt 1954). In their view, 'discontentment, disillusionment, distrust of the future – these factors helped to keep the birth rate low'.

The depopulation of Melanesia: problems with sources

Melanesia as a geographical construct

The region that has become known as 'Melanesia' was a construct of early nineteenth-century cartography, following the European recognition in the late eighteenth century that 'Polynesia' was a vast archipelago with a shared language and therefore, it was presumed, a common culture history. In 1833 the navigator Dumont d'Urville distinguished between Polynesia (the many islands), Micronesia (the small islands) in the North Pacific, and Melanesia (the black islands) for the region stretching

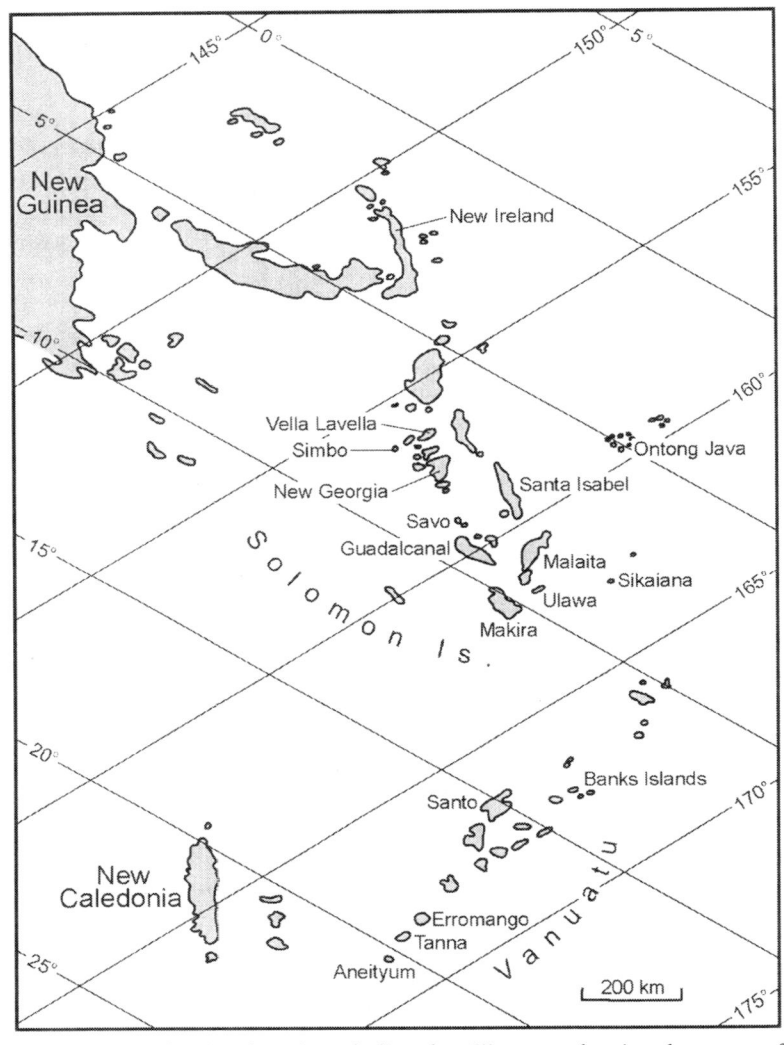

Figure 1.1. *Island Melanesia excluding the Fiji group, showing the names of islands mentioned in the text*

between New Guinea and New Caledonia (Figure 1.1). D'Urville's Melanesia originally included Australia, but this increasingly British enclave was soon separated by geographers from the rest of the black islands. Later scholars failed to find much genetic, linguistic or ethnographic evidence for a common ancestry for Melanesians, emphasising instead the region's diversity, but the name itself has survived.

A turning point in the legitimation of the name was the establishment by Anglican missionaries of the Diocese of Melanesia in 1861. Ethnologists started to adopt the term Melanesia (e.g. Rivers 1914, 1922a; Ivens 1927; Hogbin 1930), and in the post-colonial era, local elites began a discourse about the Melanesian Way. As Epeli Hau'ofa has argued, it is justifiable to continue to use the European-imposed threefold division of the Pacific "because the terms are already part of the cultural consciousness of the peoples of Oceania" (Hau'ofa 1994: 161). Furthermore, and despite the diversities of colonial rule (with British, French, German, Dutch and Australian variants), there are many common strands in recent Melanesian history. One example is the traumatic effects of initial contact between the region's indigenous peoples and the outside world, as this paper illustrates.

Problems with sources: Fiji

Melanesia's engagement with 'guns, germs and steel' did not really begin until after 1800, and the process is therefore well documented by comparison with the history of contact in the New World. Even so, the information available needs careful interpretation. In the absence of proper medical diagnoses or full and accurate population statistics, very often we are forced to rely upon the more casual observations of contemporary observers. Thomas (1994) and Jolly (1998) have reviewed some of the problems we face today in using these accounts, with Fiji as an example. The depopulation of Fiji is relatively well documented, but even here the positionality of observers (missionary, commercial or administrative) and their constructions of Fijian motherhood resulted in a biased analysis of the underlying causes of population change.

In Fiji there was a rapid decline in the indigenous population in the nineteenth century, particularly after about 1850, a trend which continued more slowly into the 1920s. For fifty years after the British colony's establishment in 1874, the state of the Fijian population was a central government concern, and mortality and fertility figures were always one of the first items mentioned in district officers' reports. This concern was used to justify widespread programmes for relocation of villages, improved sanitation, piped water and better housing. However, Thomas (1994: 112–123) points out that because the causes of population decline were poorly understood, it was actually impossible for officials to devise appropriate counter-measures.

In this situation, many policies that were justified in the name of 'sanitation' seem, in fact, to have been designed to make Fijians more visible and easier to administer, for example with the creation of larger nucleated settlements centred on the new (and usually crowded) churches and schools. This new settlement pattern and lifestyle appeared orderly and civilised to British eyes, but it also facilitated a more rapid transmission of many diseases, as the fortuitous quarantine effects of population dispersal were removed.

Without accurate theories of epidemiology and demography, the committee appointed in 1893 'to enquire into the decrease of the native population' faced a difficult task. The best that Victorian science could achieve in Fiji was listed in 36 postulated causes of the ongoing population decline. Some of these suggested factors were categorically rejected by the committee in its Report of 1896. For example, the committee discounted the alleged effects on fertility of certain marriage customs that were supposed by some observers to have had detrimental effects through in-breeding. However, many other causes could not be rejected, including many vague and speculative factors said to connect the Fijian temperament with morbidity and mortality. It was claimed that 'mental apathy, laziness and improvidence of the people arise from their climate, their diet, and their communal institutions' (Report, pp. 73–74, cited by Thomas 1994: 114). Here we see that ideological preconceptions, policy concerns and muddled thinking are so hopelessly intertwined, that even the basic 'observations' of colonial agents become suspect.

Claims for 'bad mothering' by Fijian women are a particular example (Jolly 1998). Postulated cause number 36 for depopulation, 'General insouciance of the native mind, heedlessness of mothers, and weakness of maternal instinct', was a claim supported by the testimony of high-ranking Fijian men as well as by many of the European missionaries, colonial officials and ethnographers. Some of the recommendations of the committee were for better infant diets, use of cows' and goats' milk, village crèches, encouragement of early marriage, more effective inquests into infant deaths, and deterrents against abortion. Jolly (1998: 193) points out that in relation to inquests alone, the administrative task was almost impossible. The Fijian population of 1893 was one in which 27 percent of infants died in their first month and 44 percent in their first year. But in any case, the contemporary evidence was overwhelming that exotic infections were the primary cause of this mortality, not neglect by bad mothers.

Problems with sources for western Melanesia

Elsewhere in the region, the administrative effort was more feeble, and it was mainly the missionaries who provide the primary evidence. In many cases their testimony also reflects their positionality. Florence Coombe, a teacher at the Melanesian Mission school on Norfolk Island from 1909 to 1919, provides an example at one extreme. Coombe visited numerous islands on the boat Southern Cross:

> Gaua, Banks Islands ... Everywhere are traces of a formerly large and strong population. What is left? Villages with only 30 to 50 inhabitants apiece, and amongst them not half a dozen babies. Magic and poisoned arrows have been doing destruction for generations, and sheer ignorance and laziness account for the scarcity of children. (Coombe 1911: 67).

> Guadalcanal, Solomon islands ... Everywhere the old men tell the same tale. Long ago the villages were many and large, and thickly populated. But violence and magic have mown down the people. (Coombe 1911: 331).

In point of fact, the regular visits of the Southern Cross were themselves a major cause of the transmission of infectious diseases direct from New Zealand to Melanesia (Hilliard 1978: 156–157).

The resident missionaries are generally more reliable, for example Walter Ivens who worked from 1895 to 1909 on Ulawa and south Malaita. Ivens was a brilliant linguist and an assiduous ethnographer, and the supposed effects of 'magic' form no part of his analysis. However, like all the missionaries he hated the labour trade known as 'blackbirding':

> The government of the Solomons estimates that today Mala [Malaita] has a population of about 65,000 people ... In the past the population was much greater, and the white man must be held responsible for a considerable diminution in the numbers of these people. Mala was a great recruiting ground for the vessels of the Labour Trade, and since the cessation of the Trade [in 1906], dysentery and influenza spread from visiting ships have caused considerable mortality ... The heavy mortality among children is probably owing to hookworm and yaws. (Ivens 1927: 25)

> At present [1925] the villages on Ulawa are nine in number ... The sites of deserted villages, *naonga*, are many, and the population which is at present 900 must have been a good deal larger in the past ... In 1869 there was a very severe epidemic of dysentery that swept the whole island and carried off possibly a quarter of the population. The epidemic was introduced by some ship, possibly by a vessel seeking labour ... (Ivens 1927: 44–45)

A third example is Lieutenant Boyle Somerville, of the Royal Navy, who spent eight months altogether in 1893 and 1894 surveying Marovo Lagoon, New Georgia island in the western Solomons. Somerville interacted with traders as well as local people, but surprisingly nowhere does he mention the impact of either the labour trade or disease:

> ...[I]n the eastern parts [of Marovo] the number of the population has gone down with great rapidity. An old trader of twenty years experience [Frank Wickham] told me that in his recollection the numbers had terribly decreased. This to a large extent is probably due to head-hunting which has ... almost annihilated some villages... No doubt head-hunting has always been their custom; but... rifles and especially tomahawks, during the last forty or fifty years, have largely increased its fatal effects; so that where one man's head was taken in olden times, three or more are taken today. (Somerville 1896: 410)

An uncritical reading of this contemporary literature might suggest that each island in Melanesia has its own unique history of European contact, and to some extent this may be true. Undoubtedly some islands had early and intense involvement with trading ships, blackbirders, missionaries and government agents, while other islands were more fortunate in avoiding devastating epidemics, wars or cultural malaise. However, the testimony upon which to base any assessment is patchy, often ill-informed, and always subject to ideological bias. In constructing any general model of underlying causes, these problems of data and its interpretation loom large. We need to focus on rigorous micro-studies of particular societies, where demography and/or epidemiology were analysed in some depth, in order to get beyond the dubious generalisations of an earlier age of scholarship. The islands of Simbo and Ontong Java provide such an opportunity.

The causes of depopulation in Melanesia

Recent studies of the population history of Melanesia have been rather few when compared to other parts of the world. New case studies include Fiji (McArthur 1968; Cliff and Haggett 1985; Jolly 1998), Aneityum in Vanuatu (McArthur 1978; Spriggs 1981), New Caledonia (Sand 1995, 2000), New Ireland (Scragg 1954), the Polynesian Outlier atolls (Bayliss-Smith 1974, 1975a, 1975b), and Marovo Lagoon in the Western Solomons (Bayliss-Smith *et al.* 2003). What we lack are new explanatory models which might confirm or refute the classic accounts of Rivers (1922a), Roberts (1927) and Hogbin (1930, 1939). The following sections provide a brief overview of opinions.

McArthur: how reliable are early estimates of population?

Some of the debates have been largely statistical, and concern the reliability of early population estimates. Throughout the Pacific the earliest ships usually reported the presence of many more people than were counted some years later in the first accurate enumerations. Norma McArthur (1968, 1970) was particularly sceptical about the reliability of pre-census estimates, for example those used by Harrisson (1937: 261) for Vanuatu. Harrisson proposed that a total population for the islands of 1,000,000 had collapsed to 600,000 by 1882 and 45,000 in 1935, because of new epidemic diseases. The statistical basis for the 'one million' appears not to have been published, but its origin was probably an extrapolation from the cases of Erromango and Aneityum in southern Vanuatu. It is still unclear how typical was the experience of such islands.

Despite her general scepticism about the accuracy of early estimates, McArthur (1978) showed that in Aneityum's case, the population estimates made by nineteenth-century shore-based missionaries were basically accurate, and catastrophic mortality did occur. The first reliable census on Aneityum in 1854 lists 3,800 names with about 200 additional persons not included, but even at that early date epidemic mortality had already severely reduced numbers. There are eyewitness accounts of previous epidemics of cholera and/or dysentery in about 1836–1838 and 1842–1843. These accounts suggest that in each case about 4,000 people had died, which would imply a pre-contact total in the range 9,000–12,000 (Spriggs 1997: 257–259). In 1861 one-third of the surviving population died from measles, with death rates of 40 percent for adult males and 42 percent for adult females (McArthur 1978: 278). The Aneityum population's lowest point was reached in 1941, when just 186 persons remained.

Roberts: 'centuries of decay'?

Most of these accounts, both from Aneityum and elsewhere, place primary emphasis on new sources of mortality following European contact, and this echoes the conclusions of earlier scholars. The historian Stephen Roberts (1927: 73–75), for example, summarised the historical process in the Pacific as involving three intertwined causal factors: 1) the increased scale and effects of warfare, following the introduction of firearms and other weapons; 2) the rapid and almost complete abandonment of old practices, social institutions and beliefs, leading to 'psychological despair', loss of the will to live, and neglect of children; and 3) physical causes, notably new epidemic diseases, whose mortality effects became more potent because of factor (2).

Roberts, like certain others, believed that Melanesian populations were already in decline when Europeans first arrived on the scene: 'Little wonder ... that diseases wrought such ravages on a stock already enfeebled by centuries of decay, and, at this juncture, lacking the very will to live' (Roberts 1927: 80). But to most scholars the evidence for pre-European decline was unconvincing (Hogbin 1939: 125–126). In apparent contradiction was the evidence for dense settlement and intensive cultivation, as described by Mendana in 1568 and by Quiros in 1616, on islands almost empty of people in the first nineteenth-century accounts (Spriggs 1997: 232–237). For example, the irrigated terraces and permanent villages described by Mendana on Guadalcanal cannot be traced in nineteenth-century records or in oral histories. However, one such site was recently rediscovered by archaeologists, and here post-Spanish abandonment may not have happened until the mid-nineteenth century, when escalating warfare, epidemics and blackbirding resulted in accessible areas on Guadalcanal being emptied of population and the survivors relocating their settlements in the interior (Roe 1993: 23).

Ivens: placing the Roberts model into a three-stage chronology

The Roberts model was put into a chronological framework by Ivens (1930), who wrote with particular reference to Malaita, Solomon Islands. He divided the depopulation process into three overlapping stages, beginning with the onset of labour recruitment (blackbirding) in the 1860s through to the 1890s. Ivens (1930: 42) estimated that only about half of the men recruited ever returned home from Australia, Samoa, New Caledonia or Fiji, and of those who returned, many delayed their marriages or never married at all. Therefore not only did the absentees deplete the population, but also fewer children were born than had been the case in previous years. The second stage was one of escalating warfare between the 1870s and 1890s. The introduction of rifles by traders and blackbirders was the cause of increasing bloodshed and political unrest. The third was the period of epidemic infections, between the 1880s and 1920s. The establishment of copra plantations, regular trade and the migration of large numbers into new coastal villages saw the onset of a new era of high epidemic mortality from new diseases, particularly dysentery and influenza. Ivens also observed that 'in the new villages ... which sprang up on the coast after the return of the Kanakas [recruited labourers, in 1906–1908], everyone suffered either from malaria or from ulcers [yaws], and their cemeteries are literally full of children.' (Ivens 1930: 47).

Hogbin: epidemics as the primary cause of depopulation

The social anthropologist Ian Hogbin (1939), focussing on the Solomon Islands, considered that it was mainly smallpox and dysentery which reduced the population between 1870 and 1900, in a series of epidemics:

> Since then [1900] there have also been periodic outbreaks of chicken-pox, whooping cough, measles and influenza. I have been in a native community myself during two severe epidemics, and the results on both occasions were appalling. No precautions were taken against the spread of infection, and once a person became ill his only chance of recovery was that his natural resistance would triumph, even though this was often partially destroyed through lack of proper nursing and suitable foods. Several new diseases, such as gonorrhoea, tuberculosis, dysentery, and leprosy have also become endemic. (Hogbin 1939: 127)

These new diseases had greater effects because a high proportion of the population was already debilitated by yaws, malaria and hookworm. Infant mortality rates were 'enormous, a fact which is hardly surprising when one recalls the conditions under which parturition takes place', while children's diets were 'unsuitable'.

However, Hogbin was unimpressed by the arguments of Rivers, Roberts and others that emphasised the importance of 'the psychological factor' in morbidity and mortality, as a consequence of the traumatic cultural change that followed missionisation. On the island of Malaita, for example, there seemed to be no differences between Christian and pagan areas in birth and death rates, despite the marked contrast in acculturation and, supposedly, an increased susceptibility in Christian areas to psychosomatic illness (Hogbin 1939: 132–135):

> ... [I]n all my fieldwork, not only in Ontong Java and Malaita but in other parts of Melanesia and New Guinea, I have never seen a native die of despair. Apart from accidents, all who perished were organically diseased. Sexual intercourse is also, if anything, on the increase, and everywhere the people still want children. (Hogbin 1939: 136)

Changes in fertility: of minor importance?

In almost all this literature, changes in fertility are not thought to have been of much importance. There are reports, often anecdotal, that certain customs detrimental to fertility were continuing, such as late marriage and self-induced abortion. On closer inspection, the evidence for such practices often appears unreliable (Underwood 1973). There are also reports, seldom supported by clinical diagnosis, that new STDs were reducing fertility (Pirie 1972). Usually, however, the main focus is on high epidemic mortality, and high endemic levels of infant mortality. The child death rate was being exacerbated by lack of breast milk supplements, poor hygiene, dirty clothing and bad mothering. Unless reforms could be instituted, it was suggested, Melanesians would fail to adjust to the new challenges of modernity. Their populations would not replace the numbers lost to the increased mortality rates, and extinction would become inevitable.

Depopulation and fertility: the conventional model

According to this literature, therefore, in both Melanesia and Polynesia it was changing rates of mortality that resulted in population decline, with fertility rates playing a minor role. The conventional model thus echoes the modern consensus about the causes of post-Columbian depopulation in the Americas, seeing depopulation as the result of new diseases that increased death rates, with birth rates rather little affected. In Island Melanesia (i.e. Bismarck Archipelago, Solomon Islands, Vanuatu, New Caledonia, Fiji) the chronology of the process was somewhat later than in Polynesia, and is summarised in Table 1.1. As in Australia, there were many regional variations, and the model does not apply at all to inland New Guinea where the Highlands valleys were protected from the main onslaught of epidemic disease by the late arrival of Europeans (1920s–1950s). However, in most areas the population decline occurred mainly in the nineteenth century and was driven primarily by epidemic mortality. Fertility may also have fallen in consequence, or it may have been unaffected. In either case the rising birth rates of the twentieth century were too little and too late to protect these populations from several decades of severe and sometimes catastrophic decline, although extreme cases like Aneityum may not be typical.

Table 1.1. *The depopulation of Island Melanesia: the conventional model*

Period	Demographic processes	Outcome
Pre *c.* 1840	High fertility balanced by high mortality.	Stability
c. 1840–1920	Epidemic mortality much increased; fertility somewhat reduced by death of spouses and failure to re-marry.	Rapid decline, depopulation of small islands and outlying areas
c. 1920–1960s	Control of infectious disease, acquired immunity and rising birth rates as traditional fertility checks are abandoned.	Recovery and slow increase
Post-1960s	Further reduction in incidence and effects of infectious disease; control of malaria (not applicable to Fiji).	Rapid increase

Low birth rates: the case of Makira

The interaction between epidemic mortality and the subsequent fertility of the surviving population is usually difficult to establish from historical sources. Census data are incomplete or unreliable, so that the only evidence available is anecdotal. An example is a log-book entry by the captain of a Solomon Islands copra boat, J.E. Philp:

> Wanomi Bay [Makira]. Thursday June 5th 1913 ... At noon I went across to the Marist Mission ... Later I joined the fathers at their evening meal ... We were discussing population question. Father B. [Babonneau] gave as his experience that the rate of decrease in San Cristoval [Makira] was 60 per cent! Mothers are before marriage made sterile in many cases, or again, children are unwelcome to men who wish to have the services of their women wholly at their disposal for work in gardens ... In all the villages infanticide or abortion are commonly practiced. So evidently the population must decrease. (Herr with Rood 1978: 141)

The 'evidence' for these highly suggestive assertions is, unfortunately, very slender. Philp, citing Babonneau, refers to ten mission girls from Wanomi Bay who had all been married for some years. Of these ten, seven were childless and the other two had borne one child each, of whom only one now survived (Herr with Rood 1978: 141). Not only does this information come to us second-hand, but also the sample of women is small and may be unrepresentative. But Philp appears to have been a perceptive and disinterested observer, writing his diary for no particular audience, and he too observed that in most villages on Makira there were 'few young children and many couples without issue' (Herr with Rood 1978: 142).

Low birth rates following epidemics: Fiji

We are in a better position to analyse the interactions of mortality and fertility in the case of the Fiji islands, where careful estimates of each district's population date back to 1874 and the first census took place in 1881. From these and other sources the measles epidemic of 1874 can be studied in some detail. Although McArthur (1968: 350) considered its impact 'was probably quite exceptional ... [because] conditions for its spread were extraordinarily favourable', the one-third death rate among adults was probably not unusual in smaller island populations before the twentieth century. What was 'exceptional' about Fiji in 1874 was that the epidemic spread throughout the islands, so its demographic effects became visible in the national census data for decades to come.

The impact of this extreme epidemic on the population's future viability was the result of two factors. The first was the immediate impact of high child mortality. Those infants already born in 1874 ran the double risk of dying from measles or dying from neglect because their parents and relatives were sick or dead. Contemporary observers considered that small girls were particularly at risk (McArthur 1968: 30). The second was the creation of a large number of widows and widowers whose reproductive lives were terminated or at least attenuated. It is likely that with a one-third adult mortality rate, less than half of all marriages survive an epidemic (McArthur 1968: 351), and as a result births in subsequent years are greatly reduced.

As a result of these two factors working together, the 1873–1877 age cohort in Fiji was greatly depleted in numbers. These reduced numbers, especially of girls, had an impact on birth rates 20 years later in the 1890s and 1900s, when the cohort reached marriage age (McArthur 1968: 351). Fortunately in Fiji subsequent epidemics (influenza in 1891, whooping cough also in 1891, measles again in 1903) were less severe, and in 1905, births exceeded deaths for the first time since at least 1874.

The 1874 measles disaster in Fiji affected the entire population, but a more normal pattern in Melanesia was for epidemics to be more localised, with an impact on a single island or a group of islands in frequent canoe contact with each other. If contact with the outside world resulted in these small populations experiencing epidemic mortality repeatedly, and if we assume that marriage practices did not quickly adapt, then we might predict more and more childless marriages and an accelerating decline in the number of surviving children each decade. There is indeed good evidence that supports this model from some of William Rivers' almost forgotten research in the western Solomons.

Childless marriages on Simbo and Vella Lavella

All the evidence suggests that the impact of 'guns, germs and steel' was no less severe in the Solomon Islands than elsewhere in the Pacific, but as already discussed, the process was and is poorly documented. Although the islands were declared a British Protectorate in 1893, no census took place until 1931. Population estimates for the Solomons group as a whole around 1900 ranged from 100,000 to 150,000, and all observers agreed that total numbers were in decline, with some small islands becoming depopulated (Bennett 1987: 151). The head of administration was Charles Woodford, whose first visits to the islands had been in the 1880s. In his public writings he spoke only of his fear of labour shortages in the future (e.g. BSIP 1911: 47). However, in 1910 Woodford wrote in a confidential report that 'nothing in the way of the most paternal legislation or fostering care, carried out at any expense whatever, can prevent the eventual extinction of the Melanesian race from the Pacific' (cited by Hilliard 1978: 157).

As described above, contemporary observers blamed depopulation on the combined effects of disease, labour recruiting and warfare. The social anthropologist William Rivers (1922b) was one of the few to emphasise instead the low birth rate. Rivers based his opinion on the field data that he collected with Arthur Hocart in 1908 on the islands of Tanna and Santo in the New Hebrides (Vanuatu), and in Vella Lavella and Simbo in the western Solomons. He became convinced of the importance of low fertility from his analysis of genealogies. On Simbo, these covered the entire population and were based on about four months of careful questioning and cross-checking. For the larger island of Vella Lavella, also in the western Solomons, the genealogies represent 'random samples taken from various villages of the coast', and are regarded by Rivers as 'less trustworthy'. His data from Santo and Tanna (Vanuatu) are based, in each case, on brief visits and the genealogies of single families, but interestingly these also show the same patterns (Rivers 1922b: 100–103).

The method adopted by Rivers was based on the discovery that he had made in the Torres Strait islands in 1899:

> I discovered that people preserved in their memories with great fidelity a complete and accurate record of their descent and relationships. It was possible to collect pedigrees so ample in all collateral lines that they could serve as a source of statistical enquiry into such features as the average size of family, infant mortality, and other subjects which furnish the basis for conclusions concerning the fluctuations of population. (Rivers 1922b: 96–97)

Comparing three different generations, Rivers felt confident that he could make comparisons over time in the size of families, the proportion of childless marriages, and the child mortality rate as a percentage of the total number born.

The results of his analysis are summarised in Table 1.2. The precise chronology of each generation is obviously uncertain, but when the three generations are compared some clear trends can be detected. Generation I must have been born in about the 1820s and 1830s, and were in their fertile married years between the 1840s and 1870s. In this generation there were 2.0 children per married couple, and on average this cohort of children suffered 11 percent mortality before marriage. Altogether 19 percent of all marriages were childless. Generation II were born in the 1840s–1850s, and were married and fertile in the 1860s–1890s. There were 1.3 children per married couple, and this cohort of children suffered 27 percent mortality before marriage. In Generation II 46 percent of all marriages were childless. Generation III were born in the 1860s–1870s, and were married and fertile from about the 1880s until the 1908 survey. Until 1908, there were only 0.7 children per couple and 53 percent of marriages were childless. For a few couples in this generation, their families were still incomplete at the time of the 1908 survey, so these figures slightly underestimate true fertility rates.

Rivers (1922b: 99–101) was struck mainly by the dramatic changes between Generation I and II, within cohorts where he regarded the data as fairly complete, and reasonably accurate with respect to the number of childless marriages. He was more cautious in his interpretation of the apparent rise in infant mortality:

> It is a question whether children who died young may not have been in many cases forgotten in the case of the earliest generation and therefore omitted when the pedigrees were collected, and in this case the increase in infant mortality would not be as great as

Table 1.2. *Fertility on Simbo for three generations: those born approximately 1830s and 1840s (Generation I, married and fertile c. 1850–1870); those born approximately 1850s and 1860s (Generation II, married and fertile c. 1870–1890); and those born approximately 1870s and 1880s (Generation III, married and fertile c. 1890 until 1908). Data source: Rivers (1922b: 98)*

Rivers' Generation[1]	Estimated period of fertility[2]	Marriages where the woman was of child-bearing age during the period of the Generation[3]					Total number of persons recorded for the Generation	
		Total number of marriages	Childless	1–2 children	3–5 children	6 or more children	Number of children doubtful[4]	Children born (number per marriage)[5]	Children that died before age of marriage[6]
I	c. 1850– c. 1870	207	40	90	68	9	0	407 (1.9)	44 (11%)
II	c. 1870– c. 1890	295	136	85	56	10	8	379 (1.3)	101 (27%)
III	c. 1890– 1908	110	58	36	6	0	10	72 (0.7)	33 (46%)

Notes

(1) Rivers divided his genealogies into three successive generations numbered I, II and III, Generation III being families recently completed or nearing completion in 1908. He stated that 'the division into generations was necessarily rough, but was effected before any attempt was made to estimate fertility' (Rivers 1922b: 198).

(2) Based on an estimated 20 years between generations apart from the 18 years of the incomplete Generation III.

(3) Numbers are calculated from the statistics provided by Rivers (1922b: 98) for the percentage of marriages in each category.

(4) Rivers (1922b: 95) admits that the apparent changes in birth rate between Generation II and III 'may be illusory owing to certain families [10 in the Simbo case] being still incomplete'. In contrast the eight families in Generation II where the number of children is 'doubtful' apparently reflects some gaps in the data.

(5) Rivers admits that these figures are likely to underestimate fertility: 'There is the possibility that male children who died young would be remembered better and that some female children who died in infancy may have been forgotten and therefore omitted' (Rivers 1922b: 100).

(6) Rivers (1922b: 98) describes this category as 'children who died young', but it is clear from his method of deriving this category that 'death before marriage' is what he means, although most would indeed have died as infants.

represented in the table. It will be noted that mortality is definitely greater in the case of male children, but here again there is the possibility that male children who died young would be remembered better and that some female children who died in infancy may have been forgotten and therefore omitted. (Rivers 1922b: 100)

The record collected by Rivers and Hocart from Vella Lavella, western Solomons, was similar and showed an even more serious decline in fertility. The data are a sample rather than a complete island genealogy, and 'I did not know the people and their circumstances as I knew them in Eddystone [Simbo]'. The proportion of childless marriages in Vella in Generation I was low (12 percent) and similar to Simbo, but it rose to 35 percent in Generation II (marriages undertaken between the 1870s and 1880s). The childless proportion rose to an extraordinary 72 percent in Generation III, and of the few children born between about 1890 and 1908, approximately one in four died.

Methodological problems with genealogical data

It would perhaps be unwise to submit these data to more elaborate analysis, given the approximations intrinsic to the method that Rivers adopted. Nonetheless there is now a large body of data in historical demography generated using basically similar methods of family reconstitution. An example is Hollingsworth's (1964) genealogies for the offspring of British peers in the period since 1550. The data show that between 1550 and 1699, 18 percent of all the marriages of the sons of peers (sample size 1,741) were childless, while 20 percent of their daughters' marriages (sample size 1,520) were childless (Hollingsworth 1964: 46). The proportion of childless marriages rises in some periods to 23 percent, but the true rate is never less than 16 percent (Hollingsworth 1964: 47).

Data of this kind are regarded as essentially accurate because they come from genealogies based upon well-preserved and tolerably complete historical documents, rather than ethnographic sources. In the ethnographic cases, it is always difficult to know if the genealogies recounted by informants are representative, as their only source is the testimony of families that happened to survive to the present day. In a truly random sample there should be a considerable number of families that died out, and so had no one to preserve a memory of them (Hollingsworth 1969: 211). As well as the omission of extinct families, ethnographic genealogies may also fail to represent the biological facts of descent and relatedness for other reasons:

> Informants engage in activities known to anthropologists as telescoping, clipping and patching in order to distort the biological facts so that they can be used to rationalise the social, political and economic needs and desires of the informants or the group ... These deficiencies suggest that ethnographers exercise some caution in generalising about size, sex ratios, age, and rates of growth for populations for which only ethnographic evidence is available. (Morrill and Dyke 1980: 1–2, 9)

At this remove, and because of Rivers' own early death in 1922 before his Simbo ethnography was fully published, it is difficult to know how far such comments should erode our faith in his data as a source of demographic insights. However, what is impressive about the Rivers data is that he himself exercised great care in its collection, and was cautious in its interpretation. Moreover consistent results were achieved on the two islands Simbo and Vella Lavella (and were not contradicted by sample genealogies from Vanuatu). The data deserve to be taken seriously, even if Rivers' own interpretations need to be looked at again.

The chronology of population change in western Solomons

Tentatively, therefore, we can use the Rivers data set to construct a chronology for demographic change in this region, as follows. Firstly, there is the period between about 1790 and the 1830s, when the first low-intensity contacts took place. At this time the Simbo population was still geographically dispersed, and epidemic mortality was probably small. Secondly, between about the 1830s and the 1870s there was an increasing intensity of contact, as Simbo became an entrepôt for trade and a magnet for visiting ships and beachcombers. Almost certainly many new diseases were introduced, and headhunting escalated. By 1850 bush populations were moving to the coast, but it is unclear if this was the cause or the effect of population decline. As mortality increased, the birth rate declined. And thirdly, between about the 1870s and 1910, there was a period when resident foreign traders used Simbo as a centre for copra trading, and labour recruiters made frequent visits. The population was ravaged by diseases that reduced birth rates severely. Headhunting could no longer be sustained, but was suppressed in any case by British colonial authorities after 1896. In 1903 the Methodist Mission established a toehold; and the population fell to about 400 in 1908.

This history can be pieced together from written accounts, and in 1908 it could probably have been greatly amplified by oral histories if Rivers and Hocart had not been so intent on reconstructing custom rather than studying colonial change. The Rivers-Hocart expedition took place at least 120 years after first contact, which was probably initiated with John Shortland's visit in *Alexander* in 1788. Shortland named the island Eddystone, and his sailors traded 'nails, beads and other trifles' with Simbo men who, unafraid, boarded his ship from their canoes. Ignoring invitations, Shortland decided not to land (Dureau 2001: 133–135). Shortland recommended the route via Simbo as 'the safest and most expeditious passage ... from Port Jackson to China', and within a few years more intense interactions were taking place. In 1803 the whaling ship *Patterson* recorded that 30 canoes with nearly 200 men came out from Simbo to trade (Bennett 1987: 350; pers. comm.). In 1812 Captain Bristow of the ship *Thames* reported the island was 'very thickly inhabited, which prevented them landing; it seemed full of trees, but coconuts only were obtained' (Purdy 1816: 103). In contrast, in 1908 Rivers and Hocart found there were no sea-going canoes left, and the declining population numbered only about 400 (Hocart 1922: 74).

We have no information from these brief accounts about whether or not new diseases were introduced, and no real indications of the size of the pre-contact population. However, we can assume it was relatively stable, with moderate to high fertility balanced by moderate to high mortality. Infant deaths resulted in particular from endemic malaria, while adult men died in warfare. Dureau (1998: 244) points to cultural practices on Simbo that reduced fertility, including the separate residence of men while they prepared for bonito fishing and headhunting expeditions, the breast-feeding of infants for up to four years, and the use of contraceptives, abortifacients and infanticide 'in order to maintain the ideal of two dependent children per woman'. At this remove, it is impossible to know how far this 'ideal' was actually realised.

Possibly the impact of Europeans on the Simbo population was slight until the 1830s, which is when Simbo began to gain its reputation as a safe haven for castaways. In 1839 there is a record of a white man who had left a whaling ship to live on Sikaiana atoll being transported to Simbo by another whaler (Bennett 1987: 356). The first people within Rivers' Generation I were born in the 1820s and 1830s at this time of increasingly frequent visits, as ships took advantage of Simbo's strategic location and its safe anchorage in the stormy Northwest season.

Andrew Cheyne stayed there for six weeks in 1844, curing bêche-de-mer, procuring turtle shell, and digging up sulphur from the volcano (Shineberg 1971: 303–314). The people already seemed very familiar with trade goods and trading practices, and had begun to cultivate sweet potatoes. Cheyne met three Englishmen who had been living on the island for some time, after running away from ships. He found that the Simbo chiefs and bigmen were seeking to monopolise the New Georgia trade, stockpiling turtle shell to exchange for tomahawks so they could dominate political relations with their neighbours. Cheyne left four white men on Simbo with trade goods, and later he wrote a book describing the islanders as 'on friendly terms with Europeans (which their neighbours are not)', and recommending that 'strangers should procure pilots and interpreters at the Eddystone before going to any of the other islands' (Cheyne 1852). Thirty years later this advice was still being plagiarised almost verbatim in other sailing directories (Findlay 1884: 864).

There is some confirmation of this interaction in the oral history provided to Hocart by Sulutava (Dureau 2001: 139–140). Sulutava's account indicates that the first white men to visit the island wanted turtle shell as well as bananas and coconuts. Later six white men lived on the island, each one remembered by name. Of these, three were married to Simbo women but only one of them had children. The white men took part in wars against Rendova and Marovo (Dureau 2001: 140).

It would be surprising if major new diseases had not been introduced to Simbo in this period, but the only evidence that we have is highly indirect. Hocart (1922: 76–77) recorded the recollections of a man over 70 years old in 1908, who remembered the time when the population was divided into 'bush' and 'saltwater' communities. Most people lived inland in hamlets, and these people only migrated to the coast and learnt how to fish after this informant became an adult, probably around 1845–1850. In Miller's archaeological survey at least 23 inland settlement sites were mapped as well as numerous shrines also located in places remote from the present-day coastal villages (Miller 1979: 28). Was this change in settlement pattern the cause or the effect of population decline? Cheyne (1852: 64) mentions 'fleets' of well-constructed canoes travelling from Simbo for raids up to 100 miles away, each canoe carrying 30–40 men. It is hard to imagine the 1908 population of 400 being able to man even three such vessels. It would require a total of 4,000 people on

Simbo to assemble a fleet of 30 war canoes, but unfortunately such figures are purely speculative.

Solomon Islands trade boomed in the 1870s, with regular contact with Australia by steamer and numerous labour recruiting vessels (blackbirders) and copra traders visiting the islands. When Douglas Rannie, Government Agent on the vessel *Heron*, visited Simbo in 1884 to recruit workers for the Queensland cane fields, there were already two other vessels there – the schooner *Atlantic*, trading for copra, coconut oil, pearl and turtle shell, and the *Albatross* from Fiji, also recruiting labour (Rannie 1912: 22–29). The hostility that Rannie encountered (in marked contrast to friendly welcomes in previous decades) he attributed to the large number of young men that had been 'stolen' in the recent past by blackbirders: 'When I stepped ashore I saw several hundred men emerge from the thick scrub skirting the road. They were all armed'. Of course the reference to 'several hundred men' is far from constituting a census, but his further description of 'a procession' of canoes, with 22 warriors in the leading canoe (Rannie 1912: 29, 26), does not suggest a population or society in a state of total collapse.

It is nonetheless tempting to date the beginnings of the rapid decline of the Simbo population to the 1870s, following the rise in trade and labour recruitment. Starting in this decade visiting ships made more frequent visits, contacting a population now living in nucleated coastal villages vulnerable to epidemic disease transmission and the acquisition of new endemic diseases. Starting in 1869 there were white traders resident in Gizo, and many were established on the mainland of New Georgia after 1880 and on Simbo itself after 1896 (Bennett 1987: 386). In Roviana, the Methodist leader George Brown noticed in 1899 'a great apparent decrease in the population from that which I had seen twenty years before', an impression which traders like Wickham confirmed (Brown 1908: 516).

Simbo's favoured location and resources may have made its population especially vulnerable. Bennett (1987: 365–367) has listed the cargoes of ships trading from Solomon Islands to the port of Sydney, and records 'sulphur' on seven different occasions between 1870 and 1880. There is one record of 20 tons (in 1872), two records of 15 tons (both in 1871), 6 tons in 1876, and 5 tons in 1880. Simbo, Vella Lavella and Savo are the only known sources of sulphur in Solomon Islands, but only the Simbo deposit is easily accessible (Grover 1955). The island had been a noted supplier since 1844, when Cheyne (1852: 63) loaded his ship with 'two or three tons' daily. He recommended that 'a large quantity of sulphur might easily be collected, which the natives might carry down in baskets to the stony beach, and it could be brought down to the cove in boats in fine weather'. In the 1880s, as well as the labour recruiters already described, traders were also visiting Simbo for turtle shell, pearl shell, copra and ivory nuts (Collinson 1926). It seems likely that the rapid decline in numbers dates from this period, following the introduction of new infections.

Reasons for Simbo fertility decline

The evidence suggests that Simbo's high levels of mortality and its fertility decline both date from the period after *c.* 1870, and continued up to (and beyond) the time of the Rivers-Hocart expedition of 1908. But in 1908 Rivers was surprised to find no clear evidence of epidemics of introduced disease:

> There is no record of any very severe epidemics. Tubercle and dysentery, the two most deadly diseases in Melanesia, do not appear to be, or to have been, especially active; and though both the chief forms of venereal disease exist on the island, they do not seem to have done any great amount of mischief. (Rivers 1922b: 101)

Several of the other factors commonly cited in Melanesia, including changes in clothing, house type, alcohol use and firearms, were also absent or negligible in their effects on Simbo in 1908. Because of this diagnosis, and confronted by the data on low birth rates that he generated from his genealogies, Rivers concluded that what he called 'the psychological factor' had to be invoked. Suggested effects were the reluctance of women to conceive, their eagerness to secure abortions, and their neglect of babies. By the 1870s these practices were seen as having a severe effect on population replacement, as shown by the reduced family size and high infant mortality rates. His overall conclusion has often been cited:

> We have here only another effect of the loss of interest in life which I have held to be so potent in enhancing mortality. The people say to themselves: 'Why should we bring children into the world only to work for the white man?' Measures which, before the coming of the European, were used chiefly to prevent illegitimacy have become the instrument of racial suicide. (Rivers 1922b: 104)

How far can this conclusion be trusted, when the ethnography upon it was based seems to have been relatively slight? During their fieldwork in Melanesia, neither Rivers nor Hocart seems to have had any significant interaction with female informants. Rivers himself focussed on Simbo kinship, which was the main reason for his genealogical work, as well as religious belief and sexual behaviour (Rivers 1914: 252–254, 1924, 1926). Hocart collected data on ritual, magic, ethno-medicine, and warfare (e.g. Hocart 1922, 1925, 1931). Although he was a qualified doctor, Rivers seems not to have undertaken medical diagnoses on Simbo in any systematic way, beyond treating one of his assistants for pneumonia.

In fact, there is enough published information to implicate strongly both introduced epidemics and the endemic sexually transmitted diseases (STDs) in the island's demographic history. The District Commissioner reported in 1906:

> There has been a tremendous amount of sickness among the natives, both in Simbo and Rubiana [Roviana]. They have been dying every day and are still doing so. It is carrying off all the old men and women. (Edge-Partington 1907: 22)

Rivers (1924: 32–48) states that he and Hocart were told of 'about a hundred examples of … conjoined processes of taboo and medicine'. They recorded in detail 60 cases, which included magical spells and ritual practices connected to conditions like insanity and epilepsy as well as remedies for introduced infections like pneumonia and dysentery. Epidemics were attributed to a spiritual power called Ave, whose coming was indicated by broken rainbows, shooting stars, red clouds, raindrops falling during sunshine, and also by the presence of fever, headache and cough (Rivers 1924: 47). The impression conveyed by these accounts is of a community in which much effort was invested in protection from many sources of morbidity and mortality, although perhaps this emphasis reflects in part the ethnographers' own strong interest in these topics.

The impact of STDs was not emphasised by Rivers (1922b: 101), but he reported their presence on Simbo ('both the chief forms'), and their symptoms were recorded in clinical detail by Hocart (1925: 237). These graphic descriptions are enough to demonstrate a close knowledge by Simbo men of the effects of both gonorrhoea and syphilis (S. Ulijaszek, pers. comm.). However, given the widespread prevalence among Simbo children of yaws, a disease spread by *Treponema pallidum* ssp. *pertenue*, the impact of syphilis (*T. pallidum* ssp. *pallidum*) may have been diminished by acquired immunity (Pirie 1972: 188–189). Gonorrhoea was probably the more important STD, as in New Ireland (Scragg 1954). Rivers' (1926: 71–96) own account from Simbo of sexual beliefs and practices before and after marriage indicate that STDs, if present, would have quickly spread through the unmarried population. He considered it 'exceptional and almost certainly unknown in the past' that a woman remained a virgin before marriage, and having multiple sexual partners was an accepted and integral part of a young woman's puberty rituals.

It seems unlikely that white men in the nineteenth century were in any way excluded from sexual relations with unmarried women, and as a probable outcome the widespread STD infection of men and women would have been inevitable. Rivers alludes to contraceptive and abortion practices, but his evidence consists of hearsay about rituals, spells and the ingestion of plants with unknown pharmacological properties, with no evidence at all for their efficacy. Instead, STD infection itself can account for many of the cases observed by Rivers of childlessness. It is well known from studies in West Africa that syphilis increases the rate of miscarriage, and among women who do suffer a spontaneous abortion permanent sterility often follows. In West Africa gonorrhoea also leads to sterility among both men and women (Retel-Laurentin and Benoit 1976: 291). In a rural area of Upper Volta where 31.5 percent of women had syphilis, 28 percent of women aged 50 and over were childless and 24 percent of their pregnancies had ended in miscarriages or stillbirths (Retel-Laurentin and Benoit 1976: 280). In this case the proportion of women who were childless appears very comparable to Simbo and Vella Lavella in the late nineteenth century, but whereas Rivers was inclined to blame women for securing their childlessness through induced abortion or contraception, STDs could easily have achieved the same result.

We can conclude that in the western Solomons the effects of STDs on spontaneous abortion (miscarriage) and on sterility were combined with the effects of epidemic disease on adult mortality. Such deaths resulted in many marriages being terminated by the loss of one spouse. As a result fewer children were born and inevitably there was some neglect of orphans. Finally, the high infant mortality rate among the small numbers of children born further reduced the population's capacity for replacement. The result was a decline that probably started before 1850 and accelerated in the last two decades of the nineteenth century.

Simbo after 1908

The depopulation trend on Simbo was not reversed until the mid-twentieth century. The first national census by the Protectorate government was not until 1931, and it recorded 344 persons living in five villages on the island (J.A. Bennett, pers. comm.). In addition there may have been small numbers living away from the island. It is likely that the recovery of the population did not begin until the 1930s, and only accelerated rapidly with health improvements in the 1950s. In 1960 the population had grown well in excess of 600 (Scheffler 1962: 137). Malaria eradication campaigns in the 1960s greatly reduced infant mortality levels, and by 1990 the number living on the island totalled 1,400 to 2,000 people 'depending upon the season' (Dureau 1998: 242–243). The *de jure* population of 2,000 included people living in urban or development areas, or married into other islands. With the collapse of the central Solomons state in 1999–2003, many of this diaspora will have returned home, and the current population may now exceed 2,000 people, all subsisting off the island's 12 square kilometres of land resources. It is quite possible that after 150 years the population densities of the mid-nineteenth century are again being experienced.

The depopulation of Ontong Java

A final example of the role of fertility in depopulation is provided by the atoll of Ontong Java, a Polynesian outlier located 270 km north of Santa Isabel in Solomon Islands. Ontong Java, like Simbo, is a well-documented case of the interactions between mortality and fertility, in the context of new introduced disease. This brief review will summarise evidence presented in detail elsewhere (Bayliss-Smith 1974, 1975a, 1975b), focussing in particular on the data for women's reproductive histories in the twentieth century collected by myself in 1972.

The atoll consists of two main villages, Pelau in the north and Luangiua in the south, with other smaller settlements occupied on an occasional basis. The earlier contact history of Ontong Java mirrors that of Melanesia as a whole, but the atoll's isolation, difficulties of lagoon access, poor anchorage, and organised resistance by the chiefs meant that contacts with the outside world were less intense until the 1870s. Beginning in the 1830s, there are occasional records of visits by whaling ships – for example, in 1867 the bark *Stephanie*, being short of crew after a smallpox epidemic aboard, called at the atoll and left 'with four of the natives on board' (Anon. 1867). However, until 1874 no ship entered the lagoon. In that year the *James Birnie* landed at both Luangiua and Pelau to collect and cure bêche-de-mer. Following a dispute, the ship was attacked and burnt, and many of the crew were killed (Woodford 1916: 32). In 1875 *H.M.S. Beagle* was sent on a punitive expedition and shelled Luangiua village, and in that decade visiting ships became more and more frequent. Ontong Java labour was recruited to work in German Samoa in 1878, and there are other records of blackbirder recruitment in the 1880s (Finsch 1881; Findlay 1884: 872; Wawn 1888, 1893). Copra traders also began to visit, and in 1895 a permanent trading station was established (Anon. 1895).

By about 1880, therefore, all the pre-conditions were in place for the introduction of new infections. According to Parkinson (1890), malaria first came to the atoll in the 1880s, and there are indications also of a measles epidemic. By 1889 the custom was already established of ritually cleansing the boats of visiting ships to guard against the introduction of sickness (Anon. 1890). On the other hand, visitors in the 1890s were still impressed by the size of the Ontong Java population, which was estimated to be '3,000–4,000' in 1900 and 'over 3,000' in 1901 (Admiralty 1908; Bennigsen 1901). A trader, Harold Markham, who was resident at Luangiua from 1907 to 1923, made various estimates of the population of Ontong Java at the beginning of his stay, according to second-hand reports. Of these a quoted figure of 2,500 is the most plausible, but still possibly an exaggeration (Bayliss-Smith 1974, 1975b: 427). More careful estimates in 1910 by German ethnographers, based in part on village maps, lead to an estimate of 1,300–1,400 (Sarfert and Damm 1929). The population decline after 1910 is well documented, and it continued until 1939 when just 588 people remained. Thereafter there was a recovery, gradual at first but more rapid after malaria eradication campaigns in the 1960s.

Ontong Java mortality and fertility, 1920s to 1972

We focus here on the period between the 1920s and 1972, which includes the last 20 years of actual decline. In a census of July 1921, 1,016 people were enumerated, 804 in Luangiua and 212 in Pelau. The total had fallen to 698 in 1928, when the anthropologist Ian Hogbin conducted his Ph.D. fieldwork there, and it reached its lowest ebb in July 1939 when 588 people remained (Bayliss-Smith 1975b).

Hogbin (1930) attributed the decline to an enhanced death rate that he linked to a combination of psychosomatic and pathological causes. He firmly rejected the notion that 'psychological factors' as such had any impact on mortality. However, he did suggest that the rapid transformation of traditional culture in the twentieth century, itself accelerated by depopulation, and the people's own interpretations of the supernatural origins of sickness, did lead in many cases to a fatally passive state of mind in times of ill health. These psychosomatic complications significantly increased the mortality from endemic malaria and tuberculosis, and from epidemic influenza and other exotic infections. The primary causes of decline were therefore physical, but the impact of disease was heightened by the people's 'state of mind which acquiesces in extinction' (Hogbin 1930: 65). Later accounts by medically trained officials put the emphasis exclusively on disease pathogens, in particular malaria and respiratory tract infections (e.g. Brownlees 1939; Black 1952; Hollins 1957). In 1928 four serious epidemics were remembered as having occurred since 1900 (Hogbin 1930), including influenza in 1906 which killed 30–40 persons per day (Sarfert and Damm 1929). The 1918 'Spanish' influenza pandemic reached Solomon Islands, and resulted in a suspension of labour recruiting, but its demographic impact is unknown (Bennett 1987: 176–177).

In relation to fertility decline, certain factors relevant to Simbo and other islands cannot be invoked in the Ontong Java case. There is no evidence for STDs, and Hogbin (1930, 1931a) rejected the idea of changes in sexual behaviour. In 1928 induced abortion was rare and infanticide was practically unknown. Similarly, diets were not much altered, weaning practices were unchanged, and the quality of childcare seemed unaffected by the small modifications in lifestyle induced by the copra economy. From the end of blackbirding until the 1950s, out-migration was a negligible factor.

We can nevertheless see indications that birth rates were being affected by the new sources of mortality. In 1928, out of more than 300 adult women on the atoll, Hogbin found only two cases where women had never married, but 'very frequently' families remained small because of the death of one spouse and the failure of the other to re-marry. A medical assistant stationed on the atoll for five months in 1939 counted in his census 25 widowers and 42 widows of whom many were still in their reproductive years (Kuper 1939). Childlessness was probably masked by adoption practices, but a negative impact on the birth rate can be inferred.

Introduced malaria

The role of malaria on Ontong Java also needs to be considered. The atoll's high rainfall and plentiful habitats for mosquitoes, which include *Anopheles farauti*, mean that vectors for the spread of *Plasmodium* parasites are present (Black 1952). Ian Hogbin told me in 1970 that he had never encountered worse mosquitoes anywhere, in a lifetime of fieldwork in Solomons and New Guinea, than those he experienced in Ontong Java in 1928. Once malaria was introduced, it was able to spread rapidly among people who often were living in crowded village settlements.

Extremely high rates of infection have been recorded on the atoll. In 1928 it was reported that 89 percent of the persons examined had enlarged spleens (Hogbin 1930). In 1940, 50 out of 60 children examined had palpable spleens, as did 7 percent of adults (Turner 1940). In 1952 Robert Black, a South Pacific Commission consultant, found that 'the amount of malaria has increased over the past decade in both groups of people, especially at Leuaniua [Luangiua], so that malaria is now hyperendemic and is probably approaching the holoendemic level at Leuaniua'. Black used the 1950 WHO classification, defining 'hyperendemic' as a situation where spleen rates in children were constantly over 50 percent and the spleen rate in adults was high, whereas 'holoendemic' was where the spleen rate in children was constantly over 75 percent but the spleen rate in adults was low, with strong adult tolerance of the disease. He identified *Plasmodium falciparum*, *P. vivax* and *P. malariae* at Luangiua, while in Pelau only *P. vivax* was found (Black 1952: 6–7).

As well as high infant mortality, malaria infection results in periodic attacks on the adult population. If they survive childhood infection, adults achieve semi-immunity to such attacks, but their resistance to other diseases is reduced. The malignant tertian malaria caused by *P. falciparum* is particularly deadly, and would have caused mass mortality in the virgin population of the late nineteenth century. Some idea of its potential impact can be gauged from the high rates of morbidity among American troops on Guadalcanal, and later the catastrophic mortality rates among Japanese prisoners of war (McGregor 1968). An additional effect of malaria on women is to increase greatly the incidence of miscarriages, premature births and stillbirths among women. This effect has been cited in several cases where malaria has spread to previously uninfected populations (e.g. Cilento 1928; Jones 1967; Groube 1993).

The recovery of the Ontong Java population dates from the 1940s, and can be related to the state of relative quarantine that the atoll experienced during the Second World War and the lesser incidence of epidemic infections in the post-war period. Malaria control in the 1960s greatly improved fetal and infant survival rates, and numbers began to increase rapidly. In 1970, 854 people resided on the atoll and a further 202 were living elsewhere in Solomon Islands or PNG (Bayliss-Smith 1975b: 464–466). In 1986 the resident population reached 1,408 people, and an additional 350–400 people were living in Honiara and elsewhere (Bayliss-Smith 1986: 4–10). Today's population may exceed 2,000, perhaps matching again its nineteenth-century size.

Miscarriages, stillbirths and child deaths

We can gain an indication of the demographic impact of malaria by considering the reproductive histories of 110 women aged over 30 whom I interviewed in Luangiua and Pelau in 1972. At the time of this fieldwork I had resided on Ontong Java for a total of 14 months during two periods, June 1970–May 1971 and June–August 1972. In my first period of fieldwork, I conducted a population census visiting every house, and this was updated and improved in 1972. In addition I had carried out numerous other surveys of fishing and gardening, and had become well-known to almost the entire population. Trust had been further improved by my wife's role as teacher in the primary school. Few of the women spoke pidgin and I conducted my interviews with them using interpreters (David Kaia'enga in Luangiua, Patterson Pohiu at Pelau), but by 1972 my command of the Ontong Java language was sufficient for me to understand much of the dialogue. The information that women supplied to me was often confirmed (sometimes corrected) by others in the household. None of the topics discussed seemed to be regarded as either private or particularly sensitive, apart from the obvious sadness that these women felt in recalling the many deaths and disappointments during their childbearing years.

For simplicity's sake, the following analysis will ignore some differences between the two atoll communities, Luangiua and Pelau, which are explored in detail elsewhere (Bayliss-Smith 1975b). The 110 women who were interviewed represented a 95 percent sample of all women in these age categories living on the atoll in 1972. The sample included only three who reported no pregnancies at all in their lives. Two others had been pregnant at least once but were childless because of miscarriages, infant deaths and/or widowhood. Unlike on Simbo, the proportion of childless marriages on Ontong Java does not seem to have been abnormal in the period studied, whereas the somewhat low frequencies of pregnancy and the high levels of fetal and infant mortality were much more significant.

Inter-pregnancy intervals

The mean age of marriage for women in the sample was 19 years. I obtained reliable information on the onset of menopause from 30 older women, but four of these women said they had stopped menstruating before the age of 35 years, for unknown reasons. Excluding these four individuals generates an average age of 42 years for menopause. Since very few pregnancies occurred before marriage, the data suggest a potential reproductive period of about 22 years, which in practice was reduced by early widowhood to an actual average of 18.2 years for the 76 post-menopausal women in the sample (Bayliss-Smith 1975b: 453). How many pregnancies did these women experience in their period of potential childbearing?

In total, 792 pregnancies were reported giving rise to 803 zygotes (there were 11 twins) and to 692 live births (Table 1.3). The sample is sub-divided by time period in Table 1.3, which also shows a calculated mean inter-pregnancy interval and a mean inter-birth interval. In both cases these intervals have shortened since the 1920s, and particularly in the most recent period 1961–1972. Between the early 1920s and 1940 many pregnancies (at least 30 percent) resulted in miscarriage, stillbirth or death of the infant within a week. Such events might have permitted a woman to become pregnant again quickly, with an inter-pregnancy interval of 12 months or less. Despite this potential for

Table 1.3. *The reproductive histories of 110 Ontong Java women aged over 30 years in 1972: mean intervals between perceived pregnancies and between live births, for four different time periods, 1920s–1972*

Period	Perceived pregnancies Sample size	Mean interval (months)	Live births Sample size	Mean interval (months)
1921–40	108	31.4	91	37.4
1941–50	186	31.0	166	33.7
1951–60	236	29.0	217	32.6
1961–72	262	25.7	218	30.7
Entire period	792	28.7	692	32.9

Source: Bayliss-Smith (1975b: 440–459), with slight revisions in the light of new information collected in 1986. The intervals were calculated by counting the number of pregnancies/live births that occurred in each time period, and then dividing this figure into the total number of reproductive years available. Reproductive years were counted as the period during which each woman was both married and below the age of menopause, either reported (in 30 cases) or assumed (in 46 cases, based on 42 years which was the average reported age of menopause – see text).

reduction, the mean inter-pregnancy interval was prolonged, being 31.4 months before 1940 and 31.0 months from 1941 to 1950. These intervals are close to the physiological maximum that has been observed in populations with extended lactation and almost no infant mortality (Wrigley 1969: 92). As nutritional stress is unconvincing as an explanation – the atoll diet is rich in coconut fat and fish protein – we therefore need to invoke cultural explanations for the delayed onset of pregnancy.

Very few women travelled away from the atoll until the 1960s, and customs surrounding marriage, sex and childbirth probably remained unchanged until at least the 1950s. The Anglican Mission did not achieve its first success on Ontong Java until 1936, and few girls attended even primary school until the 1960s. We can therefore envisage a persistence of cultural practices that separated spouses for long periods, for example women's residence for mourning in the cemeteries, and ritualised avoidance of sexual intercourse after the first childbirth (Hogbin 1931a, 1931b; Bayliss-Smith 1975b). Probably the effects of these practices and of prolonged breast-feeding only began to diminish in the 1950s. By 1961–1972 the mean inter-pregnancy interval had fallen to 25.7 months, despite an improvement in fetal and infant mortality rates that might have resulted in the opposite trend.

Mortality rates of birth cohorts

The percent mortality of cohorts originating in different time periods is shown in Table 1.4. Fetal mortality is necessarily an estimate based on the women's own perception of a pregnancy. As in all societies, early miscarriages are either unrecognised or are forgotten, and the data from Ontong Java undoubtedly omit them. In many cases the sex of a spontaneously aborted fetus was known to the mother and was reported to me, indicating that late miscarriages were mainly what I recorded. We might also expect some under-reporting because of forgetfulness by some older women. In Table 1.4, stillbirths are added to the category of miscarriages. Taken together they account for 17 percent (at least) of all pregnancies in the period before 1940. Infant deaths occurring between birth and the age of 5 years account for another 19 percent loss, with further deaths occurring in this cohort before it reaches the age of 20 years. This measure, mortality from conception (as perceived by the mother) until age 20 years, remains very high (between 39 and 44 percent) until the end of the 1950s.

The more conventional mortality measure, deaths from birth to age 20 years, removed more than 30 percent of all those in cohorts born from 1920 to 1960. While not as catastrophic as the

Table 1.4. *Ontong Java cohorts for perceived pregnancies and for live births, showing mortality up to age of 20, by time period, 1921–1972*

Period	Number reported Pregnancies	Number reported Zygotes	Miscarriage or stillbirth	Birth to age 4	Age 4 to age 19	Conception to age 19	Birth to age 19
1921–1940	108	109	18	21	8	43	32
1941–1950	186	190	24	39	12	39	31
1951–1960	236	241	24	70	13	44	38
1961–1972	262	263	45	32	10	33	19
Entire period	792	803	111	162	43	39	30

Source: as Table 3. My census of 1986 enabled the mortality of the 1960–1972 birth cohort to be updated to include deaths in the years 1972–1986. However, as those individuals born between 1966 and 1972 had not yet reached the age of 20 in 1986, the mortality rate given for the age group 5–19 may be a slight underestimate. However, rather few deaths occur in late teenage years, especially under the conditions of the late 1980s, so the mortality rate shown is probably very close to being correct.

mortality reported on Simbo and Vella Lavella by Rivers (see Table 1.2), such losses greatly erode the possibilities for demographic recovery. Only after 1960 does the situation begin to improve. If such infant and child mortality rates were characteristic of the period between about 1880 to 1920, and if epidemic mortality was also striking down adults in significant numbers thus terminating marriages and cutting the numbers of pregnancies, then the rapid depopulation of Ontong Java over this period can be explained readily

Undoubtedly it is malaria's impact that is mainly responsible for both fetal and infant mortality levels on the scale indicated by the Ontong Java data. It is impossible now to reconstruct fully the history of malaria in Solomon Islands, but it is clear that population movements within the colonial state did much to spread new forms of the disease to populations previously quarantined or semi-immune. New nucleated villages, meeting houses, churches and schools also did much to facilitate transmission of the parasite to susceptible individuals, once new *Plasmodium* species had been introduced. Perhaps even the arrival of new strains of existing *Plasmodium* species could overcome the locally-acquired semi-immunity of adults, and thus produce new epidemics (Groube 1993).

On Ontong Java the demographic recovery of the 1940s took place at a time of enforced isolation because of the Second World War and its aftermath. All epidemic infections were much less severe, even though the existing malaria infection had become endemic. Respiratory infections and measles were absent or less prevalent in the 1950s, and in the 1960s malaria control began. The World Health Organization malaria control programme became fully effective in 1970, spraying houses with DDT and at the same time treating malaria cases with chloroquine. In the 1960s the mortality of the birth cohort was halved, at a time when inter-pregnancy intervals were falling with new customs affecting marriage and sexual intercourse. Ontong Java's traumatic colonial experience with 'guns, germs and steel' was finally at an end.

Conclusion

The two case studies have shown that where micro-scale data are available, more complex explanations for depopulation must be constructed than those assumed by most contemporary observers. Both Simbo (syphilis and gonorrhoea) and Ontong Java (malaria) demonstrate the subtle and hidden effects of the new infections. By affecting age structures and terminating reproductive lives through widowhood, epidemics have insidious effects on fertility as well as on death rates. Through high rates of sterility, fetal loss and infant mortality, such diseases in turn reduce birth rates. The cases of Simbo and Ontong Java also illustrate the difficulties faced by weak colonial authorities in countering depopulation, when so much of the discourse surrounding the question was based on ignorance, bias and muddled thinking.

To avoid the simplistic generalisations of an earlier age of scholarship, we need to study the depopulation of Melanesia through a nuanced reading of the historical sources, better epidemiology, and an explicit analysis of the interactions between fertility and mortality. There are implications for the historical demography of the Americas, where epidemic mortality is often assumed to be a sufficient explanation for depopulation, in a region lacking the wide range of information sources that we have available in Melanesia. Guns and Steel were important aspects of colonialism, but the role of Germs is more subtle than is usually assumed, and in Melanesia it is a factor needing much further research.

Acknowledgements

For valuable information about Simbo, I am much indebted to Judy Bennett (University of Otago) and Edvard Hviding (University of Bergen). For his assistance with the retrospective diagnosis of STDs, I thank Stanley Ulijaszek (University of Oxford). For discussions on demography I am grateful to Vern Carroll (Michigan), Richard M. Smith and Janice Stargardt (Cambridge), and also Les Groube (Le Bourg, Brittany) whose ongoing work on malaria has been a particular inspiration. The late Ian Hogbin was also generous with advice at an early stage in my Ontong Java fieldwork. None of my data collection there could have been achieved without the help of the people of Luangiua and Pelau, and in particular the expertise and friendship of the council secretary David Kaia'enga, who acted as my interpreter in both 1970–1972 and 1986.

References

Admiralty 1908. *Pacific Islands, Volume 1 (Western Groups). Sailing Directions*, Admiralty Hydrographic Office, London.
Anon. 1867. 'Journal of the Bark "Stephanie" of New Bedford, James G. Sinclair, Master, 1864–68', Pacific Manuscripts Bureau, Canberra, microfilm 221: 62–214.

Anon. 1890. 'Auszug aus einem Berichte S.M. Kreuzerkorvette 'Alexandrine', betreffend den Besuch der Lord Howe-gruppe (auch Ontong Java genant), westlich der Salomon Inseln', *Mitteilungen aus den Deutschen Schutzgebieten* 3: 87–88.

Anon. 1895. 'Statistik', Nachtrichten über Kaiser-Wilhelms-Land und den Bismarck-Archipel 11: 41.

Bayliss-Smith, T.P. 1974. 'Constraints on Population Growth: the Case of the Polynesian Atolls in the Pre-contact Period', *Human Ecology* 2: 259–296.

_____. 1975a. 'The Central Polynesian Outlier Populations since European Contact'. In *Pacific Atoll Populations*, ed. V. Carroll, 286–343. Honolulu: University of Hawai'i Press.

_____. 1975b. 'Ontong Java: Depopulation and Repopulation'. In *Pacific Atoll Populations*, ed. V. Carroll, 417–484. Honolulu: University of Hawai'i Press.

_____. 1986. *Ontong Java Atoll: Population, Economy and Society, 1970–1986*, Occasional Paper 9, South Pacific Smallholder Project, University of New England, Armidale.

_____, Hviding, E. and Whitmore, T.C. 2003. 'Rainforest Disturbance and Histories of Human Disturbance', *Ambio* 32(5): 246–352.

Bennett, J.A. 1987. *Wealth of the Solomons. A History of a Pacific Archipelago, 1800–1978*, Honolulu: University of Hawai'i Press.

Bennigsen, V. 1901. 'Über eine Reise nach den deutschen und englischen Salomons-Inseln', *Deutsches Kolonialblatt* 12: 113–117.

Berndt, R.M. and Berndt, C.H. 1954. *Arnhem Land – its History and its Peoples*, Cheshire. Melbourne.

Black, R.H. 1952. 'Malaria in the British Solomon Islands', *SPC Technical Paper* 33, South Pacific Commission, Noumea.

Blaut, J. 1993. *The Colonizer's Model of the World. Geographical Diffusionism and Eurocentric History*. New York: Guilford Press.

Brown, G. 1908. *George Brown, D.D., Pioneer-Missionary and Explorer. An Autobiography*. London: Hodder & Stoughton.

Brownlees, J.K. 1939. 'Office of the District Commissioner, Ysabel, 28th July 1939, to the Secretary to the Government, Tulagi', in *History, Notes and Records of Lord Howe (Ontong-Java), 1900–1952*, ms. file in District Office, Auki, Malaita, Solomon Islands.

BSIP 1911. *Handbook of the British Solomon Islands Protectorate*, Government of the BSIP, Tulagi, Solomon Islands, 1–47.

Cheyne, A. 1852. *A Description of Islands in the Western Pacific Ocean North and South of the Equator with Sailing Directions, together with their Productions, Manners, and Customs of the Natives; and Vocabularies of their Various Languages*. London: J.D. Potter.

Cilento, R.W. 1928. *The Causes of Depopulation in the Western Islands of the Territory of New Guinea*. Canberra: Government Printer.

Cliff, A.D. and Haggett, P. 1985. *The Spread of Measles in Fiji and the Pacific. Spatial Components in the Transmission of Epidemic Waves through Island Communities*, Department of Human Geography Publication HG/18, Research School of Pacific Studies, Canberra: Australian National University.

Collinson, C.W. 1926. *Life and Laughter 'Midst the Cannibals*. London: Hurst & Blackett.

Coombe, F. 1911. *Islands of Enchantment: Many-sided Melanesia*. London: Macmillan.

Crosby, A.W. 1993. *Germs, Seeds and Animals. Studies in Ecological History*. Armonk, New York: M.E. Sharpe.

Diamond, J.M. 1997. *Guns, Germs and Steel: the Fates of Human Societies*. London: Jonathan Cape.

Dureau, C. 1998. 'From Sisters to Wives: Changing Contexts of Modernity on Simbo, Western Solomon Islands'. In *Maternities and Modernities. Colonial and Postcolonial Experiences in Asia and the Pacific*, eds K. Ram and M. Jolly, 239–274. Cambridge: Cambridge University Press.

———. 2001. 'Recounting and Remembering "First Contact" on Simbo'. In *Cultural Memory: Reconfiguring History and Identity in the Postcolonial Pacific*, ed. J.M. Mageo, 130–162. Honolulu: University of Hawai'i Press.

Edge-Partington, T.W. 1907. 'Ingava, Chief of Rubiana, Solomon Islands: Died 1906. Extract from a Letter', *Man* 7: 22–23.

Findlay, A.G. 1884. *A Directory for the Navigation of the South Pacific Ocean*, 5th edn. London: Holmes Laurie.

Finsch, O. 1881. 'Bemerkungen über einige Eingeborne des Atoll Ongtong-Java ("Njua")', *Zeitschrift für Ethnologie* 13: 110–114.

Flannery, T.F. 1995. *The Future Eaters. An Ecological History of the Australasian Lands and Peoples*. Sydney: Reed.

Groube, L. 1993. 'Contradictions and Malaria in Melanesian and Australian Prehistory'. In *A Community of Culture: the People and Prehistory of the Pacific*, ed. M. Spriggs et al., 164–186. Canberra: Department of Prehistory, Research School of Pacific Studies, Australian National University.

Grover, J.C. 1955. 'Simbo Volcano'. In *Geology, Mineral Deposits and Prospects of Mining Development in the British Solomon Islands Protectorate*, ed. J.C. Grover, 46–50. London: Western Pacific High Commission, Honiara, and Crown Agents.

Harrisson, T. 1937. *Savage Civilization*. New York: Alfred A. Knopf.

Hau'ofa, E. 1994. 'Our Sea of Islands', *The Contemporary Pacific. A Journal of Island Affairs* 6: 148–161.

Herr, R.A. with Rood, E.A. 1978. *A Solomons Sojourn: J.E. Philp's Log of the Makira 1912–1913*. Hobart: Tasmanian Historical Research Association.

Hilliard, D. 1978. *God's Gentlemen. A History of the Melanesian Mission, 1849–1942*. St Lucia: University of Queensland Press.

Hocart, A.M. 1922. 'The Cult of the Dead in Eddystone of the Solomons, Parts 1 and 2', *Journal of the Royal Anthropological Institute of Great Britain and Ireland* 52: 71–112, 259–305.

———. 1925. 'Medicine and Witchcraft in Eddystone of the Solomons', *Journal of the Royal Anthropological Institute of Great Britain and Ireland* 55: 229–270.

_____. 1931. 'Warfare in Eddystone of the Solomon Islands', *Journal of the Royal Anthropological Institute of Great Britain and Ireland* 61: 301–324.

Hogbin, H.I.P. 1930. 'The Problem of Depopulation in Melanesia as Applied to Ongtong Java (Solomon Islands)', *Journal of the Polynesian Society* 39: 43–66.

_____. 1931a. 'The Sexual Life of the Natives of Ontong Java (Solomon Islands)', *Journal of the Polynesian Society* 40: 23–34.

_____. 1931b. 'The Social Organisation of Ontong Java', *Oceania* 1: 399–425.

_____. 1939. *Experiments in Civilization: the Effects of European Culture on a Native Community of the Solomon Islands*. London: George Routledge.

Hollingsworth, T.H. 1964. 'Demography of the British Peerage', *Population Studies* 18(2), suppl., 1–108.

_____. 1969. *Historical Demography*. London: Hodder & Stoughton.

Hollins, F.R. 1957. 'The Atoll of Ontong Java, Its Depopulation and Malaria Control', *Journal of Tropical Medicine and Hygiene* 60: 231–237.

Ivens, W.G. 1927. *Melanesians of the South-East Solomon Islands*. London: Kegan Paul, Trench, Trubner & Co.

_____. 1930. *The Island Builders of the Pacific*. London: Seeley Service.

Jolly, M. 1998. 'Other Mothers: Maternal "Insouciance" and the Depopulation Debate in Fiji and Vanuatu, 1890–1930'. In *Maternities and Modernities. Colonial and Postcolonial Experiences in Asia and the Pacific*, eds K. Ram and M. Jolly, 177–212. Cambridge: Cambridge University Press.

Jones, L.W. 1967. 'The Decline and Recovery of the Murut Tribe of Sabah', *Population Studies* 21: 133–157.

Kuper, G. 1939. 'Tulagi Hospital, Tulagi to the Acting Senior Medical Officer, Tulagi Hospital, 20th July 1939'. In *History, Notes and Records of Lord Howe (Ontong Java), 1900–1952*, ms. file in District Office, Auki, Malaita, Solomon Islands.

McArthur, N. 1968. *Island Populations of the Pacific*. Canberra: Australian National University Press.

_____. 1970. 'The Demography of Primitive Populations', *Science* 167: 1097–1101.

_____. 1978. 'And Behold, the Plague was Begun among the People'. In *The Changing Pacific. Essays in Honour of H.E. Maude*, ed. N. Gunson, 273–284. Melbourne: Oxford University Press.

McGrath, A. 1987. *'Born in the Cattle': Aborigines in Cattle Country*. Sydney: Allen & Unwin.

McGregor, J.D. 1968. *Malaria in the Island Territories of the Southwest Pacific*. Honiara: Government Printing Office.

Miller, D. 1979. *Report of the National Sites Survey 1976–1978*. Honiara: Solomon Islands National Museum.

Morrill, W.T. and B. Dyke 1980. 'Ethnographic and Documentary Demography'. In *Genealogical Demography*, eds W.T. Morrill and B. Dyke, 1–9. New York: Academic Press.

Parkinson, R. 1890. 'Aussterben auf den Atolls in der Südsee', *Mitteilungen Geographischen Gesellschaft für Erdkunde* 1890: 197–204.

Pirie, P. 1972. 'The Effects of Treponematosis and Gonorrhoea on the Populations of the Pacific Islands', *Human Biology in Oceania* 1(3): 187–206.

Purdy, J. 1816. *Tables of the Positions ... Composed to Accompany the 'Oriental Navigator', or Sailing Directions for the East Indies, China, Australia etc. with Notes, Explanatory and Descriptive*. London: Whittle and Holmes Laurie.

Rannie, D. 1912. *My Adventures among South Sea Cannibals. An Account of the Experiences and Adventures of a Government Official among the Natives of Oceania*. London: Seeley Service.

Retel-Laurentin, A. and D. Benoit. 1976. 'Infant Mortality and Birth Intervals', *Population Studies* 30: 279–293.

Reynolds, H. 2001. *Indelible Stain? The Question of Genocide in Australia's History*. Ringwood, NSW: Penguin.

Rivers, W.H.R. 1914. *The History of Melanesian Society*, 2 vols. Cambridge: Cambridge University Press.

———, ed. 1922a. *Essays on the Depopulation of Melanesia*. Cambridge: Cambridge University Press.

———. 1922b. 'The Psychological Factor'. In *Essays on the Depopulation of Melanesia*, ed. W.H.R. Rivers, 4–113. Cambridge: Cambridge University Press.

———. 1924. *Medicine, Magic, and Religion*. London: Kegan Paul, Trench, Trubner & Co.

———. 1926. 'Sexual Relations and Marriage in Eddystone Island of the Solomons'. In *Psychology and Ethnology*, ed. G. Elliot Smith, 71–94. London: Kegan Paul, Trench, Trubner & Co.

Roberts, S.H. 1927. *Population Problems in the Pacific*, Routledge, London.

Roe, D. 1993. 'Prehistory without Pots: Prehistoric Settlement and Economy of North-West Guadalcanal, Solomon Islands', Ph.D. thesis, Australian National University, Canberra.

Sand, C. 1995. *'Le Temps d'Avant': La Préhistoire de la Nouvelle-Calédonie. Contributions à l'Etude des Modalités d'Adaptation et d'Evolution des Sociétés Océaniennes dans un Archipel du Sud de la Mélanésie*, L'Harmattan, Paris.

———. 2000. 'Reconstructing "Traditional" Kanak Society in New Caledonia: the Role of Archaeology in the Study of European Contact'. In *The Archaeology of Difference. Negotiating Cross-Cultural Engagements in Oceania*, eds R. Torrence and A. Clarke, 51–78. London/New York: Routledge.

Sarfert, E. and H. Damm. 1929. *Luangiua und Nukumanu. 1. Allgemeiner Teil und materielle Kultur. Ergebnisse der Südsee-Expedition 1908–1910*, series 2, subseries B, vol. 12, part 1, series ed. G.Thilenius, Friedrichsen, De Gruyter, Hamburg.

Scheffler, H.W. 1962. 'Kindred and Kin Groups in Simbo Island Social Structure', *Ethnology* 1: 135–157.

Scragg, R.F.R. 1954. 'Depopulation in New Ireland: a Study of Demography and Fertility', MD thesis, University of Adelaide. Reprinted 1954 by Administration of Papua and New Guinea, Port Moresby.

Shineberg, D., ed. 1971. *The Trading Voyages of Andrew Cheyne, 1841–1844*. Canberra: Australian National University Press.

Somerville, B.T. 1896. 'Ethnographical Notes in New Georgia, Solomon Islands', *Journal of the Anthropological Institute of Great Britain and Ireland* 26: 357–412.

Spriggs, M. 1981. 'Vegetable Kingdoms: Taro Irrigation and Pacific Prehistory'. Ph.D. thesis, Australian National University, Canberra, published by Department of Prehistory, ANU, and University Microfilms, 1986, Ann Arbor, Michigan.

Spriggs, M. 1997. *The Island Melanesians*. Oxford: Blackwell.

Stannard, D.E. 1989. *Before the Horror. The Population of Hawai'i on the Eve of Western Contact*. Honolulu: University of Hawai'i Press.

Thomas, N. 1994. *Colonialism's Culture. Anthropology, Travel and Government*. Cambridge: Polity Press.

Turner, J.L. 1940. 'Medical Officer to the Senior Medical Officer, 1st March 1940'. In *History, Notes and Records of Lord Howe (Ontong Java), 1900–1952*, ms. file in District Office, Auki, Malaita, Solomon Islands.

Underwood, J.H. 1973. 'The Demography of a Myth: Abortion in Yap', *Human Biology in Oceania* 2(2): 115–127.

Wawn, W.T. 1888. 'Private Logs of Cruises in the Pacific Islands, 1888–1900', ms. in Mitchell Library, Sydney.

———. 1893. *The South Sea Islanders and the Queensland Labour Trade*. London: Swan Sonnerschern.

Woodford, C.M. 1916. 'On Some Little-known Polynesian Settlements in the Neighbourhood of the Solomon Islands', *Journal of the Royal Geographical Society* 48: 26–54.

Wrigley, E.A. 1969. *Population and History*. London: Weidenfeld & Nicholson.

CHAPTER 2

THE IMPACTS OF COLONIALISM ON HEALTH AND FERTILITY: WESTERN NEW BRITAIN 1884–1940

C. Gosden

Approaches to the understanding of colonialism have shifted over the past 20 years, mainly under the impact of post-colonial theory. Earlier views tended towards a 'fatal impact' theory (named after a book by Alan Moorehead, 1966) in which native races of the Pacific were seen as unable to resist the strength of arms, the organisation and desire for riches of European nations. The fatal impact theory contained a notion of blame, but also had implicit within it a view that the strongest would always prevail. What better measure of strength is there than population size? Impacts of Europeans were literally fatal as demonstrated by declines in population throughout the Pacific region. Post-colonial theorists find such ideas demeaning to the colonised as they only accord agency to the colonisers, so that the original inhabitants can only bear the impact of the outside power as best they can (Bhabha 1994, Said 1978, 1993, Guha 1988).

Post-colonial theory emphasises local differences in colonial cultures and not monolith colonial structures deriving from the desires of the colonists. Local differences in colonialism arose because of the impact of local cultures, which nuanced, changed and subverted the aims and desires of outsiders. A common model for many influenced by post-colonial theory is that of the joint colonial culture, which is created by the social and culture logics of both locals and incomers, becoming something different from either. Colonialism creates new forms of difference out of older cultural forms, mixing and recombining cultural elements to bring to life something quite new. Such views have been developed for the Pacific by Thomas (1991, 1994) and Gosden and Knowles (2001) and for North America by White (1991) who usefully coined the term 'the middle ground' in describing the forms of life that emerged through the meeting of fur traders and Algonquian Indians in the Great Lakes region. Views emphasising the middle ground do seem the most realistic manner to describe cultural encounters in New Guinea. However, where such views are at their weakest is in accounting for demographic declines due to introduced diseases. In a review of colonialism over the last 5,000 years, I have made the point that it is only in the last 500 years that colonial forms have caused massive demographic changes first through mass death of the indigenous inhabitants of the Americas, the Pacific and Australia, with a later replacement by people of European descent in the Americas, Australia and New Zealand to form the settler societies which have such a dominant impact on the world today (Gosden 2004).

Papua New Guinea was settled late in the modern colonial period, at the end of the nineteenth century, and never saw an influx of large numbers of Europeans to create a settler society, despite some early colonial hopes. However, coastal Papua New Guinea did suffer considerable population declines between the end of the nineteenth century and middle of the twentieth, with populations rising steadily since then. Probably because of the potential overtones of a fatal impact theory, population decline and its effects have not been highlighted in recent views of colonialism. My aim in this chapter is to address the implications that population declines have for our models of colonialism. People in western New Britain were profoundly affected by sudden mass death, as might well be expected, but both population numbers and cultural forms have bounced back in a manner profoundly influenced by the need to deal with population declines.

Populations and colonialism

Colonialism is a phenomenon dependent on population density. Incoming colonials were attracted to areas of densest population, as these provided sources of labour, trade goods and food. Within

New Britain, initial European settlement took place on the Gazelle Peninsula partly due to its central position and good harbours, but more especially because of the density of people in the region. The initial desire for trade at the end of the nineteenth century gave way to a plantation economy at the beginning of the twentieth and it was then that the desire for labourers grew especially strong. Furthermore, colonial activities caused the further clustering of indigenous people around plantations, trading posts and mission stations, so that colonial centres represented a resource that local people were keen to use.

Colonialism might be seen as mutually resourcing. Labour and trade were the basis of the new colonial economy, but plantations and missions could be used as sources of new mass-produced goods or as a new market to sell food and locally-made artefacts. Colonialism caused much re-ordering of the population, as we shall see. However, the shadow cast by mass death prejudiced the whole of the colonial endeavour by reducing the numbers of people who could work and trade. Death on a large scale also posed questions of morality for all sides. For local people, colonialism was a profoundly challenging phenomenon and was seen as setting a series of moral problems. Crucial to these was the question of why local religious and ritual structures had failed to provide both the large amounts of goods enjoyed by whites, and to protect people's physical and spiritual well-being. Death on a large scale through disease and violence not only put at risk the continuance of the group, but also challenged its cosmological basis.

By the time that colonial regimes were set up in Melanesia, liberal whites thought that part of their task was to ensure the well-being of the peoples they ruled. The decline in native populations from the nineteenth century onwards caused much discussion and attempts to seek the causes so that these could be remedied. Rivers (1922), writing a final essay in a book he edited entitled *Essays on the Depopulation of Melanesia*, ran through the various physical factors causing population decline. These included death through epidemics, a lowering of the birth rate due to venereal disease, and the dislocations caused by the labour trade. However, he focused ultimately on what he called the 'psychological factor' which he felt was taking away the will to live and leading in the end to 'racial suicide'. Central to psychological health was religion. 'The old life of the people was permeated through and through by interests of a religious kind, based on a profound belief in continued existence after death and in the influence of the dead upon the welfare of the living' (Rivers 1922: 111). Study of religious institutions was necessary to understand their role in life as a whole and to create syncretisms between incoming Christianity and local beliefs, which allowed the logic of the latter to influence the manner in which Christianity was practised. Issues of morality and of morale were joined in Rivers analysis, with the ultimate thought that it would reflect badly on the civilising endeavours of the colonials if depopulation continued to its final conclusion. Rivers was not alone in his concerns about population. The government anthropologist for New Guinea, E.W. Chinnery, gathered much data on population decline during the 1920s and 1930s, with the aim of writing a major study on the issue, which, however, never eventuated.

Population histories, densities and declines lie at the heart of the colonial process and of our understanding of colonialism in the present. Let us now turn to see how these factors have worked themselves out in the case of western New Britain.

Population history in colonial West New Britain

New Britain is a large island just south of the equator with generally sparse populations at its western end, especially in the inland areas. Linguistic divisions and histories provide a means of understanding some of the diversity and unity of the society in western New Britain prior to colonialism (Figure 2.1). As Ross writes (1988: 160) no published descriptions exist for most of the languages of this area. Nevertheless, some tentative groupings can be made. The Lamogai chain (also known as Bibling (Chowning 1996)) is made up of the Mouk, Aria and Lamogai languages (Gimi, Ivanga and Lamogai); the Passismanua chain contains Miu, Kaulong and Sengseng with Psohoh at a slightly further remove; the West Arawe chain is composed of Arove (Arawe), Aiklep (Agerlep) and Apalik (Palik); the East Arawe chain is made up of Akolet, Avau, Atiu and Bebeli (Kapore); with Mangseng being an outlier of the larger Arawe chain (Ross 1988: fig. 5).

The groups that I am particularly concerned with in this chapter are the Passismanua and the East and West Arawe chains, with the Arawe area as the main focus. However, I shall draw on the broader context of New Britain as a whole in outlining colonial history. The term 'Arawe' used later in relation to material culture (Gosden and Knowles 2001) is one of the three main 'culture areas' within West New Britain discerned by Chowning (1978), each with its own set of material culture. This is covered by Arawe, Bibling and Passismanua (Chowning's Whiteman) speakers on the south coast and adjacent inland areas (here called

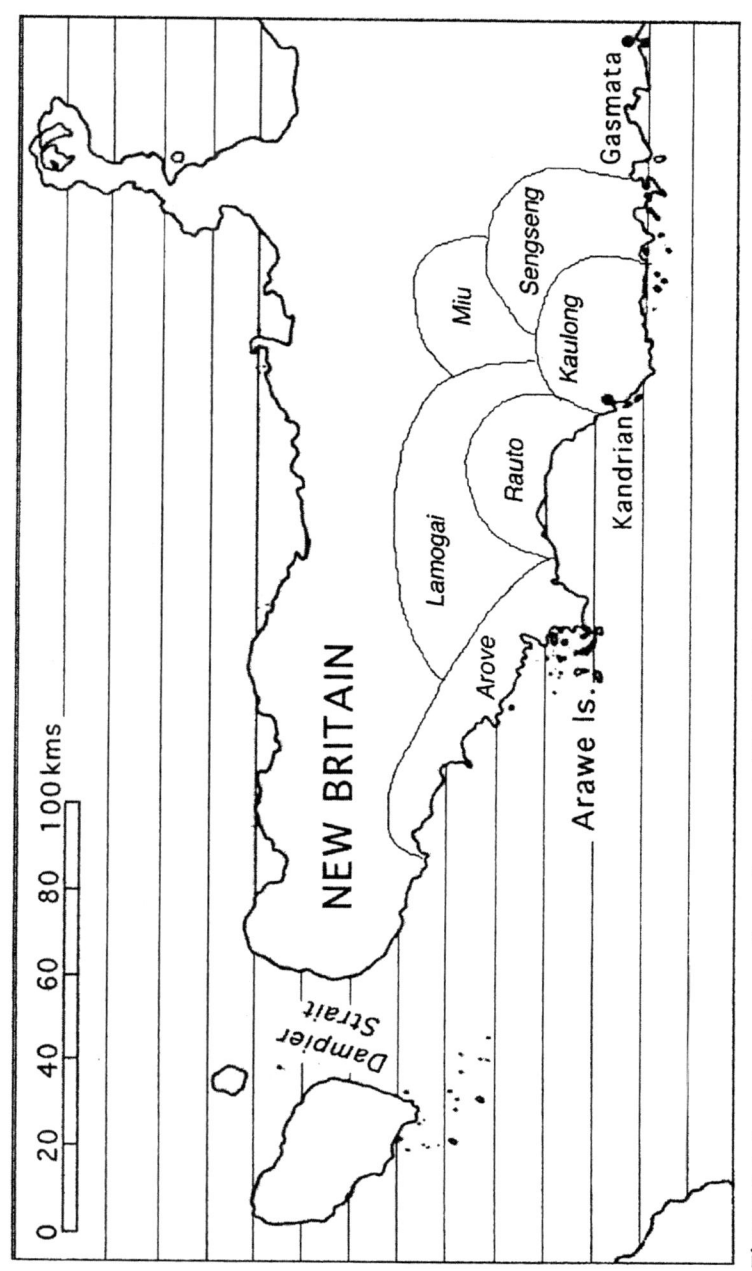

Figure 2.1. Language groups on the southern coast of West New Britain

Arawe as a shorthand designation). This area is characterized by skull deformation, blow pipes, mokmok and singa stones (used as wealth and for sorcery), pearl shells, three-piece shields, bark-cloth belts for men, and rounded pig-tusk mouth ornaments. Most of this material is still in use and is vital for initiations, bride-price and mortuary payments, as well as in exchange.

The colonial period started officially in 1884 with the annexation of New Guinea by Germany, but little direct impact was felt prior to the start of the twentieth century, with the major exception of the impact of diseases as detailed below. Between 1900 and 1914 the colonial economy was transformed from that of trade to plantations, which had a series of effects on the local population. New Guinea became 'a copra colony', although trade still supplied more than 50 per cent of the Protectorate's copra exports in 1909 (Firth 1973: 134, 136). The new plantations alienated land and there was a series of wars on the Gazelle over land from the late 1890s to early in the twentieth century. One result of this was the setting up of native reserves of land which could not be further alienated. However, much of this land might have been used for the plantations indirectly. The new plantations had a considerable demand for food. Some of this was imported but a good deal was grown locally. The very generous allowances of the Forsayth plantations (it was the Forsayths who set up the first plantation at Ralum in 1886) gave each worker 680 g of rice per day, plus 2500 g of fresh fruit and vegetables per day and 400 g of fish or meat a week (either fresh or imported) (Firth 1973: 213). All the rice was imported, but most of the fruit and vegetables were produced locally, as was a proportion of the meat and fish. By 1913, at the peak of the plantation economy, 14,990 labourers were employed in the Bismarck Archipelago as a whole (Firth 1977: 15). At a very rough estimate, this would provide a demand for 37,000 kg of fresh fruit and vegetables per day, plus whatever contributions of fish and meat were made from the local economy. This would represent a considerable investment of time and labour, and considerable areas of land would have been brought under cultivation to supply the plantations. However, it was noted that people of the Sulka and Mengen areas were re-settling near the Gazelle Peninsula in order to garden for the plantations (Sack and Clark 1979).

The income from the sale of food would have been important and have helped induct people into a monetary economy. Steel axes and bush knives, in demand at this time, would have helped clear larger areas for planting. Thus when plantations were set up on the south coast of New Britain around the turn of the century, similar sets of relationships would have pertained, and it is no surprise that there were considerable shifts in settlement pattern and subsistence at about this time. On the Arawe Islands where people were previously spread among a number of small hamlets on the south coast of New Britain and a number of islands, they now clustered in big villages on five islands and one area of the mainland (Gosden and Pavlides 1994). In the present, people in many parts of the province, and particularly on the coast, live in villages of several hundred inhabitants, but these were often formed often around the beginning of the twentieth century from a number of smaller hamlets, each of which was defended and centred around a men's house (Counts and Counts 1970: 92; Zelenietz and Grant 1986: 204). Chinnery (1925, 1926) has reported an identical form of organization for the Kaulong and other areas of the south coast.

Throughout western New Britain there seems to have been a single form of settlement pattern and social structure, which might be quite ancient, dating back around 1,000 years (Gosden and Pavlides 1994). It allowed for both mobility of settlement and open links between groups. What happened from the early twentieth century onwards is that the settlement pattern changed from small hamlets to large villages, but the open network of connections continued and expanded. People congregated into larger villages, but were still able to move regularly from one area to another on trading expeditions or to take up gardening land in a new place through kin connections. New larger settlements formed a labour pool for larger gardens to feed the settlements themselves, but also the new needs of plantations and traders for food.

The new villages were also focal points in the movement of people through the area utilizing their kin connections to travel far and wide. The large villages existed at first only on the coast, where government influence was greatest, settlement agglomeration happening inland later. All of these changes brought into question the nature of the community. Further change was brought about by the numbers of people (mainly men) leaving villages to work on plantations or in towns, drawn by the activities of labour recruiters, as they were euphemistically called. Recruiting around New Britain, even at its western end, was intense. Between 1912 and the start of 1914 there was a ban on recruitment in the Sulka and Mengen areas and the whole coastline from Cape Gloucester to Montagu Harbour. It was noted that ruthless recruiting by the Forsayth plantation had made villagers flee inland. One of their recruiters, Karl Münster, was sentenced to three months in jail, imprisonment of recruiters being an extremely rare act (Firth 1973: 224, 228). In 1913, a Forsayth recruiter broke the prohibition by obtaining labourers for the company's Arawe plantation from Rauto on the adjacent coast. In the same year, recruitment was reaching its limit in coastal areas: it is estimated that nearly every unmarried man in the villages of the northwest coast of New Britain was a recruit that year or had been one recently (Firth 1973: 173), and much the same might have been true of the south coast.

The other set of devastating changes that came about in the German period were epidemics of disease. Smallpox and influenza epidemics spread through New Guinea in 1893–1894 and had major effects in the Bismarck Archipelago (Parkinson 1999 [1907]: 90–91). Estimates of how many people died are difficult to come by. In Witu, plantations were set up on land left vacant after 50 percent of the population died from a combination of smallpox and the lethal suppression of two uprisings in 1901 and 1903 (Firth 1973: 136). Witu was densely populated so might have suffered more than some areas from density-dependent infectious diseases. However, the Arawe Islands also represent high levels of population, and according to local testimony many died at this time. A further devastating event which had a major effect on the west end of New Britain and its south coast was the collapse of the Ritter volcano in 1888. Ritter is in the Vitiaz Strait and its collapse created tsunamis which wiped out villages in the Cape Gloucester area, along the coast of Umboi and along the south coast, where local people have stories of waves several metres high. Loss of life would have been high on Gloucester and Umboi, but less on the south coast where destruction of villages and gardens would have been the main effects. This geological event, essentially random when viewed in terms of historical process, would have added an extra dimension to the confusion and disruption of the times.

Many of the plantations had been set up in the German period and went into Australian hands through the activities of the Expropriation Board. They represented small mixed communities living in close proximity. The Expropriation Board's description of Arawe Plantation, published in 1925, gives a total size of 606 hectares, carrying 51,070 coconuts and being worked by 170 labourers. They were housed in eight sets of married quarters, plus ten labourers' houses, all made of local materials. It seems that not all labourers could have lived in these houses, and thus quite a number must have commuted from villages within the Arawe Islands. There was one Chinese dwelling and a trade store, plus kerosene, rice and drug stores, and a hospital. The bungalow occupied by the European manager was made of weatherboard and was the one building made of non-local materials, and had a separate bathroom, pantry, kitchen and office. In addition, there were four copra dryers, a wharf and a wharf shed, a 3.5 ton cutter, 5 horses, 50 wild pigs, 78 goats and 17 fowl. The Military Terrain Handbook (No. 57, 1943) contains an aerial photograph (1943: fig. 15) showing the layout and indicating that not much has changed since 1925. We have no details on the community, but presumably it was made up of a European, a Chinese and workers from different parts of Melanesia.

The history of mission activity on the south coast of New Britain is obscure and difficult to disentangle from the various records which exist. There seems to have been no formal missionary activity on the south coast of New Britain during the German period or in the immediate post-war years. But between the late 1920s and 1935, both Anglican and Catholic missions were established in a number of areas along the south coast. The missions had a complex set of effects, which, like their history, is hard to sum up succinctly. At a socio-economic level they would have acted like plantations, in that they caused the alienation of small but important areas of land from local hands. They would also have needed locally grown food to supplement foodstuffs coming from Rabaul. Food would have been exchanged for material items or money, introducing an extra source of Western goods. This brief and rather prosaic overview of the changes from the late nineteenth century into the twentieth, hides the magnitude of what happened to local people. In European terms, the changes represent a combination of the Black Death and the Industrial Revolution rolled into one. The epidemics of the 1880s and 1890s killed large numbers of people, possibly in the order of 60-80 percent, if we can extrapolate from places like Bali-Witu. Older and younger people were probably most at risk, so that both the transmission of historical knowledge and the biological replacement of the community were put at risk.

The shift from a large number of smaller hamlets to fewer large villages is a complex one, but would have entailed technical changes, such as the adoption of metal technologies for clearing new large gardens and the acceptance of new crops to the region, especially sweet potato which came to complement taro. Novel social arrangements would have been necessary as clan-based smaller settlements combined into larger villages, with implications for trade and ceremony. The continuance and increase of labour recruiting at the start of the twentieth century imposed an extra social strain, as many fit young men and some women, were taken out of the community for a number of years at a time. Levels of recruiting were still a cause for concern in the 1950s in this region, and were mentioned regularly in patrol reports, although by this time populations were starting to increase at a considerable rate, accelerating into the present century. All of these factors placed the nature of the community at issue. No one was sure what the community should consist of and how it should be regulated. Part of the means of regulation was through creative work with ritual structures and forms of ceremony. In a study of changes in material culture in the Arawe region, Chantal Knowles and I studied four different museum collections dating between 1910 (made by A.B. Lewis) and 1937 (made by Beatrice Blackwood), complemented by our understanding of what was in use in the 1980s and 1990s (Gosden and Knowles 2001). One thing that struck us is how little many objects had changed. However, this was not always the case and a major decline can be seen in the availability and collection of objects associated with warfare, stone tools, containers and material culture pertaining to music.

To take stone tools first: people in New Britain adopted steel as soon as it became available, substituting stone tools with axes, knives, and razor blades, probably in that order. These changes in technology had major ramifications for gardening, as steel axes made it easier to clear larger areas of garden and this in itself might have been a response to the demands of the plantation managers for fresh food. Steel axes also made it easier to make planks for houses, and this suited the demands of the colonial administration for larger and more settled villages. It would seem, therefore, that settlement patterns, gardening and stone tool technologies changed rapidly and without fuss, contrary to functionalist archaeological expectations that these constitute large and difficult transformations (Binford 1983).

Warfare and music were probably linked in a rather unexpected fashion, as both were elements of an exclusively male domain of life in fighting and ritual (where much music was practised). Lewis arrived in 1909 at the end of the old order on the south coast of New Britain; there was still much fighting, and although some large villages were to be found, there were also smaller defended settlements located away from the coast. At that time male prestige depended partly on prowess in warfare, and there was a rigid separation of men from women and children. The men's house was the centre of male life and of active male-only cults. As colonial peace was imposed and more young men were removed from communities to work on plantations, the focus of male life shifted, and with it relations within the community as a whole. What came to replace war and male prestige was an emphasis on exchange. In the pre-colonial period exchange had been an important part of life and it was a means by which a person could gain prestige, but it was limited by the fear of travelling around a landscape in which warfare was endemic (Harding 1967). Warfare, as we have seen, was the basis of indigenous male prestige at the time, and the enforced cessation of warfare by the colonial authorities became, paradoxically, the basis of colonial prestige: pacification was how colonial regime took control. Once this avenue for male prestige had been cut off, and along with it the emphasis on exclusively male activities, exchange expanded to fill the gap. The increase in the number of different objects is paralleled by an increase in the range of items in circulation. This is difficult to quantify, but it is certain that there has been an inflation in certain types of payment, such as bride price, with far more objects needed to complete a bride price now than earlier in the twentieth century (Gosden and Pavlides 1994: table 2).

Many objects are used in a variety of forms of ritual, such as initiation or death rituals, and it is likely that payments of all types have increased. Such rituals and forms of exchange, involving the kin group as a whole, are now central to people's lives and have replaced the male-centred rituals of the nineteenth century. We feel that it is no coincidence that most such rituals concern life cycles and initiation of children into the group. Such an emphasis on the group is partly as a result of the many changes to the nature of group life, through alterations of settlement patterns and movements of people as a result of labour recruitment, as well as shifts in the relationships between men and women. The group as a whole has had to be re-thought by the people composing it, and they have used existing cultural means, constructed in novel ways, to carry out this re-thinking. The objects that have dropped out of use and circulation, apart from stone tools, are mainly to do with war and male-only ritual. As the nature of community became an increasingly important issue, rituals concerning the community as a whole increased, and circumcision, previously only the realm of men, came to be conducted in the open spaces in the middle of villages, rather than hidden in the men's house. Individuals were linked to the community, and as a man's ability to define himself through war vanished, the skills of the big man and woman, who could create and maintain large personal networks, came to be valued. Both individual and community are, of course, difficult terms, but people were individuated in new ways through the existence of larger communities, and forms of communalism were made anew to include whites through the exchange of objects ending up in European collections and the supply of food to plantations, as two prominent examples.

People also experimented with ceremony. From the beginning of the twentieth century, if not before, cargo cults became a periodic, but recurrent, feature of life in western New Britain. Lattas (1998) has documented the occurrences of cargo cults in New Britain as a response to colonialism generally and various forms of missionary activity more particularly, especially the recent work of the New Tribes Mission. Lattas's argument is that the adoption of Christianity may, in part, have been seen as a strategic route to obtaining western goods, so that Christianity was used as a resource by local people in a similar manner to that which missionaries saw local people as a set of possibilities for salvation. From the 1930s onwards, both Catholic and Anglican missions were established on the south coast of New Britain, resulting in mass conversion, as well as competition between the two churches. As elsewhere in PNG, Christianity represented an addition to existing religious practices and not a replacement of them, so that now, local ceremony coexists with the introduced faith. Coincidentally or not, Christianity was introduced at about the time when populations started to increase again, an increase that became marked after the Second World War. What is cause and effect here is hard to judge.

Conclusions

At the beginning of the twenty-first century there is again a population problem, but now one of rapid increase, not decline. This is success of a sort, partly due to the effective blending of Christianity and local beliefs. Health takes many forms and the physical and spiritual are intimately linked, as Rivers realised long ago. Population, in turn, is an excellent barometer of health, but is also key to the changes brought about by colonialism. Population decline, especially where it is dramatic, as was the case in New Britain in the 1880s and 1890s, might seem to confirm the fatal impact hypothesis. Rivers, Chinnery and others were worried that this would be the case.

My argument is rather different. The biological threat of disease to groups with no immunity to it brought on a set of creative responses of quite unprecedented nature. People on the south coast of New Britain re-organised their lives physically, living in new villages, with novel crops and technology. They also addressed the threat to the community directly by developing aspects of ceremony most relevant to the nature of the community as a whole, with emphasis placed on entry into the group through initiation rites, changes of state such as marriage, and exit from the community through death. The shocks of colonialism, physical and mental, have brought about creative responses by all concerned. Health and well-being in the broadest senses are key to these responses.

Acknowledgement

I am very grateful to Stanley Ulijaszek both for his patience and his editing skills.

References

Bhabha, H. 1994. *The Location of Culture*. London: Routledge.
Binford, L. 1983. *In Pursuit of the Past*. London: Thames and Hudson.
Chinnery, E.W.P. 1925. 'Notes of the Natives of Certain Villages of the Mandated Territory of New Guinea', *Territory of New Guinea, Anthropological Reports* Nos. 1 and 2. Melbourne: Government Printer.
_____. 1926. 'Certain Natives of South New Britain and Dampier Straits', Territory of New Guinea, *Anthropological Reports* No 3. Melbourne: Government Printer.
Chowning, A. 1978. 'Changes in West New Britain Trading Systems in the Twentieth Century', *Mankind* 11: 296–307.
_____. 1996. 'Relations among Languages of West New Britain: an Assessment of Recent Theories and Evidence'. In *Studies in Languages of New Britain and New Ireland, vol. 1: Austronesian Languages of the North New Guinea Cluster in Northwestern New Britain*, ed. M.D. Ross, 7–62, Pacific Linguistics, Series C-135. Canberra: Department of Linguistics, Research School of Pacific Studies, Australian National University.
Counts, D.E.A. and D. Counts 1970. 'The Vula of Kaliai: a Primitive Currency with Commercial Use', *Oceania* 41: 90–105.
Firth, S.G. 1973. 'German Recruitment and Employment of Labourers in the West Pacific before the First World War'. Unpublished D.Phil. thesis, University of Oxford.
_____. 1977. 'German Firms in the Pacific Islands 1857–1914'. In *Germany in the Pacific and the Far East 1870–1914*, eds J.A. Moses and P.M. Kennedy, 3–25. Brisbane: University of Queensland Press.

Gosden, C. (2004). *Archaeology and Colonialism. Culture Contact from 5000 BC to the Present*. Cambridge: Cambridge University Press.
_____ and C. Knowles 2001. *Collecting Colonialism. Material Culture and Colonial Change*. Oxford: Berg.
_____ and C. Pavlides 1994. 'Are Islands Insular? Landscape vs. Seascape in the Case of the Arawe Islands, Papua New Guinea', *Archaeology in Oceania* 29: 162–171.
Guha, R. (ed.) 1988. *Selected Subaltern Studies*. Oxford: Oxford University Press.
Harding, T.G. 1967. *The Voyagers of the Vitiaz Strait*. American Ethnological Society Monograph 44. Seattle: University of Washington Press.
Lattas, A. 1998. *Cultures of Secrecy: Inventing Race in Bush Kaliai Cargo Cults*. Madison: University of Wisconsin Press.
Moorehead, A. 1966. *The Fatal Impact : an Account of the Invasion of the South Pacific, 1767–1840*. London: Hamilton.
Parkinson, R. 1999 [1907]. *Thirty Years in the South Seas*, trans. J. Dennison. Bathurst: Crawford House Press.
Rivers, W.H.R. (ed.) 1922. *Essays on the Depopulation of Melanesia*. Cambridge: Cambridge University Press.
Ross, M. 1988. *Proto Oceanic and the Austronesian Languages of Western Melanesia*, Pacific Linguistics, Series C-98. Canberra: Department of Linguistics, Research School of Pacific Studies, Australian National University.
Sack, P. and Clark, D. (eds) 1979. *German New Guinea: the Annual Reports*. Canberra: Australian National University Press.
Said, E. 1978. *Orientalism*. London: Routledge and Kegan Paul.
_____. 1993. *Culture and Imperialism*. Random House: New York.
Spivak, G. 1987. *In other Worlds: Essays in Cultural Politics*. London: Routledge.
Thomas, N. 1991. *Entangled Objects: Exchange, Material Culture and Colonialism in the Pacific*. Cambridge, Mass.: Harvard University Press.
_____. 1994. *Colonialism's Culture: Anthropology, Travel and Government*. Oxford: Polity Press.
White, R. 1991. *The Middle Ground. Indians, Empires, and Republics in the Great Lakes Region, 1650–1815*. Cambridge: Cambridge University Press.
Zelenietz, M. and Grant, J. 1986. 'The Problem with Pisins', Parts I and II, *Oceania* 56: 199–214, 264–274.

CHAPTER 3

PURARI POPULATION DECLINE AND RESURGENCE ACROSS THE TWENTIETH CENTURY

Stanley J. Ulijaszek

As elsewhere in coastal and island New Guinea, population in the Purari delta underwent severe decline after European contact, showing resurgence only across the second half of the twentieth century. While population processes in New Guinea after about 1950 are reasonably mapped, understanding of population change at earlier times remains poor. In this chapter, patterns of population change in one region of lowland Papua, the Purari delta, are described from early colonial times to the year 1996, and possible reasons for the decline and subsequent resurgence examined. It is argued that the population decline in the early colonial period could largely be attributed to infectious diseases introduced with the contact experience, and to low crude birth rate associated with the recruitment of adult males for plantation labour. It is also argued that the population increase in the later colonial period and after independence in 1975 may be attributed to the introduction and increased availability of biomedical practice, and improved nutrition. It is suggested that the latter factor has lead to increased fecundity of women, while both factors have lead to greater survivorship of young children. High total fertility rates (TFR) and population increase post-independence is examined in relation to economic change and modernization using demographic data collected by the author in 1995 and 1997. Multiple regression analysis shows that young child mortality and maternal income level are important factors supporting the maintenance of high fertility, while maternal education is associated with reduced TFR. It is also argued that one characteristic of the continued use of palm sago for subsistence and as an income-generating commodity has allowed economic change without reduced survivorship of young children.

The Purari: land and population

The Purari population comprises six tribes (in F.E. Williams (1924) terminology), the Baroi, I'ai, Kaimari, Koriki, Maipua and Vaimuru. The term 'tribe' is problematic as a group descriptor for the Purari. However, in the absence of an alternative appropriate descriptor, I will use this term throughout this chapter, but acknowledging its imperfection in the present context. The land the Purari population inhabits is predominantly tidally-inundated swamp, heavily vegetated by nypa and mangrove, and is extremely marginal for agriculture. The subsistence system of the Purari people, as observed in the early 1980s, is well-suited to the physical realities of the Purari delta (Ulijaszek and Poraituk 1983), relying at its core on palm sago, with hunting, the use of small bush gardens and fishing providing dietary diversity. Palm sago is a staple foodstuff whose production by the Purari people is linked to demography, at least among the Baroi tribe, in that sago-gardens may be inherited at marriage and new sago gardens are often planted when children are born. However, while members of the Ia'i tribe do not appear to plant sago palm now, they are clear that their grandparents did (J. Bell, pers. comm.).

There is archaeological evidence for sago subsistence near the Purari delta, at about 1,500 years ago (Rhoads 1982). Sago-use is likely to have undergone various changes before the present pattern of sago subsistence was established in the Purari delta. At its simplest, the exploitation of wild sago palms allows the maintenance of sago grounds for continued use as part of one type of hunter-gatherer foraging practice. This may have been how sago was used by the Purari when the interior of this region was initially colonized by humans. The archaeological record also shows the Purari population to have cultivated palm sago for trade for perhaps 300 years (Hope *et al.* 1983), indicating that surpluses relative to subsistence needs could have been generated easily at that time, at least on a periodic basis. In areas adjacent to the Purari, habitation or creation of selected non-swamp areas allowed local intensive, if geographically dispersed, gardening, in addition to sago-making. This was a practice that the Purari are suggested to have copied during early colonial times (Williams 1924), leading to the garden-clearing and sago-cultivation mix observed by Hipsley and Clements (1950) and Ulijaszek and Poraituk (1983).

While warfare among adjacent clans may have been a central principle around which fortressed settlements were organized prior to colonial administration, with pacification, the peoples of the Purari delta organised themselves in settled villages (Maher 1961) on sites associated with availability of gardening grounds, as well as proximity to fishing, crab collecting and sago-making areas (Ulijaszek and Poraituk 1983). Sago-making also became one means by which the Purari population engaged with economic modernization, initially through the Tom Kabu movement of 1946–1955, generating sago surpluses for sale in the capital, Port Moresby. The I'ai village, Mapaio, where the Tom Kabu movement started, was moved to its present location because of both proximity to sago stands, and ease of transport out of the Purari delta. The involvement of other villages in this movement included relocation to sites more favourable for cash-cropping.

Population and fertility prior to European colonization

Evidence of population size and process prior to European colonization can only be inferred from a very limited archaeological record in adjacent sago-using areas. The abundance of palm sago lead Knauft (1993) to conclude that lowland Papua has always had the potential to support relatively large population aggregations. This starchy staple is easily stored and has high energy density when dry. It can therefore: (1) buffer populations against short-term food insecurity; (2) support large-scale feasting; and (3) sustain travellers for prolonged periods. The one deficiency of palm sago as a staple food is its lack of nutrients, apart from carbohydrate (Ulijaszek 1983). However, it is likely that plentiful stands of sago could have provided a staple food which could be supplemented by abundant fish and other riverine resources, given that the efficiency of sago production, in terms of time spent to generate a million kilocalories of food energy, is similar to that of shifting cultivation and higher than that of sedentary agriculture (Ulijaszek and Poraituk 1983). If nutrition from these other sources were adequate, high fertility schedules may have been the norm, if food supplies could ensure adequate nutritional status for women to have high fecundity. However, a factor which would have tempered the tendency to high fertility schedules for the women would have been extensive breastfeeding. If prolonged breastfeeding took place, as elsewhere in New Guinea, then this would have provided immunological protection against infectious diseases, reducing potential mortality among the infants, as well as inhibiting the fecundity of the mother (Ellison 2003). Although early introduction of weaning foods can disrupt this relationship, this practice was very uncommon in the Purari delta in the 1950s (Lyn Calvert, pers. comm.), and it seems unlikely that the Purari populations would have undertaken early dietary supplementation in earlier times. Despite all efforts to promote early dietary supplementation by maternal and child health services from Kapuna Hospital, the practice has not gained widespread acceptance.

High death rates would have kept the population in check by way of diseases which carry high young child mortality. Malaria, which was endemic in the Purari delta prior to colonial contact (Ashford and Babona 1980) is likely to have been particularly important in this respect. Malaria during pregnancy is associated with high likelihood of low birth weight and greatly increased risk of death in the child, for a range of reasons as a consequence of this. Malaria in childhood is associated with high mortality, independently of birth weight. Another disease category which may have had impact on population size and structure prior to European colonization is diarrhoea, which carries high young child mortality in the contemporary developing world, usually in synergy with undernutrition (Ulijaszek 1990). It is not known if nutritional stress predisposed the populations of the Purari delta to diarrhoeal infection. However, the formation of nucleated settlements for defence purposes in relation to inter-clan warfare is likely to have increased the possibility of transmission of density-dependent diseases, including diarrhoea and respiratory infections. Thus, the disease profiles of these fortress villages may have been more similar to those of sedentary agriculturalists, than of foragers, even prior to the appearance of European colonizers.

Fertility and reproduction in the early colonial period

Durrad (1922) suggests that population decline in Melanesia began with first contact with Europeans. There is no evidence to support or refute this view in the Purari delta. However, population estimates across the period of colonial administration suggests a dramatic decline, from about 20,000 people in 1907–1908 (Annual Report for Papua 1907–1908), to about 8,000 in 1920–1924, and about 5,000 in the 1950s (Maher 1961). The value for 1907/1908 is based on observation, not on enumeration. However, the Annual Report for Papua, 1907–1908 states that:

> There are six villages in the Delta – Maipua, Kaimari, Iai, Kairu, Koropenaira, and Ukiaravi. These villages, as the resident Magistrate says, probably contain 20,000 inhabitants. When the government went to Ukiaravi, between 50 and 60 canoes of Kaimari assembled to give them a 'send-off'; the largest canoes held 22, and the smallest 10 fighting men, the large canoes were more numerous than the small, so that there must have been quite 800 men in all, exclusive of those who remained in the village and took no part in the display. Ukiaravi and Iai are both more populous than Kaimari and the others are all large villages, so that 20,000 is by no means an exaggerated estimate.

While the reliability of this estimate may be questioned, the population decline across the first half of the twentieth century was real enough. The prime reason put forward for this reduction by Maher (1961) is a decline in crude birth rate (as a possible consequence of Rivers' (1922) psychological factor), which in 1950–1955 stood at 26 per thousand of population. However, this value was about twice that of the United Kingdom in the same period, and the importance of high mortality due to infectious disease cannot be ignored. The crude death rate for the Purari population across the period 1954–1979 was calculated by Hall (1983) as 19 per thousand of population. This was based totally on deaths at, and reported to, Kapuna hospital, and Hall (1983) acknowledges that inevitably this is a gross underestimate. Given that mortality rates began to decline soon after Kapuna hospital was established, it is likely that the crude death rate in the period 1950–1955 would have been similar to, or in excess of, the crude birth rate at that time. Hall (1983) postulates that the most likely reason for population decline in the early twentieth century would have been the introduction of epidemic infectious diseases to a susceptible population. There is support for both views of population decline. The breakdown of modes of cultural reproduction (Maher 1961), and the out-migration of males to oilfields, plantations, and to the Public Works Department in Port Moresby (Oram 1992) could all have contributed to low fertility. The introduction of new infectious diseases to which the Purari were susceptible included pneumonia and measles, both of which would have incurred significant mortality, and venereal disease, introduced to workers from the region who went to work for oil exploration companies, which is associated with sterility (Hall 1983).

The loss of interest in life described by Rivers (1922) for Melanesia more generally may have had some impact on fertility, but is likely to have operated only very indirectly. British colonialism suppressed inter-clan warfare, head-hunting, and cult activities. These rigorously enforced changes largely obliterated pre-colonial forms of cultural reproduction, changing Purari society forever (Maher 1961). The break-up and dispersal of large village groups associated with these changes removed many of the basic integrating forces of village community (Knauft 1993). Reduced fertility would have been one consequence of this; however, the dispersal of fortress villages with the suppression of inter-clan warfare may have served to reduce the transmission of some density-dependent diseases, and the mortality that went with them.

Pacification of the region by colonial administration allowed increasing entry of outside influence. The Purari had begun to work for wages as early as 1913 (Oram 1992). However, such influence was slow to take hold, by 1920 being limited to visits by government officers, some missionaries, and a number of traders and planters whose enterprises did not persist (Knauft 1993). A palm-alcohol enterprise was initiated in 1920, only to close in 1923, and a saw mill and trade store were established in 1922 (Maher 1961). More importantly, European influence was spread in the Purari delta by the increasing attractiveness to adult Purari males of wage labour outside the region. Most adult males left the region for work for periods of one or two years, but some left permanently. Migration levels in the Purari were extensive, the majority of young men being customarily absent from their villages, contributing to the erosion of village structure, with associated implications for fertility and population. Durrad (1922) commented that recruitment of plantation labour had been a great source of depopulation in rural Melanesia from the beginning of colonial times. The British recruiters in early colonial times were allowed to recruit women for domestic service only, although many British planters disregarded colonial regulations and took women for labour. With such out-migration, subsistence production is likely to have suffered, and fertility rates are likely to have suffered a decline for a combination of reasons. These include reductions in marriage rates and late marriage because of male absenteeism, and reduced fecundity because of a decline in subsistence production and nutritional state. Polygyny, although common, is unlikely to have been promoted by the low availability of male marriage partners across this period; rather, the influence of Christianity served to reduce the importance of polygyny in Purari society (Maher 1961).

By the Second World War, wage labour had been broadly accepted across the Purari villages. There was a significant shift from recruitment to voluntary migration for work. Some efforts at locally-initiated private enterprise had also been attempted in the Purari delta, including the establishment of a local sawmill. Although these were not so important economically as the wage labour experience had been, they were both contacts with the European economic system that were allowed to develop after the colonial suppression of warfare.

While the lowered population densities of settlements may have reduced the possibility of transmission of existing density-dependent diseases, new diseases introduced by the European colonizers, as well as by increased internal migration of adult males for work, may well have more than compensated for this reduction (Denoon 1989). For example, pneumonia was responsible for extremely high mortality rates in early colonial times, especially among indentured labourers (Riley et al. 1992). Measles is likely to have been introduced with the earliest European contact in New Guinea. This spread beyond the reaches of administration, affecting populations living at high density, including the Purari, before European contact in the region. Serological surveys of measles antibodies in various populations in East and West New Guinea between 1959 and 1960 have shown that only the remotest groups in West Papua had not previously been exposed to this disease (Adels and Gajdusek 1963). The critical population size for measles to persist in a community is estimated at between 250,000 and 400,000 (Keeling and Grenfell 1997), making epidemics of this disease extremely sporadic in New Guinea, but with high mortality when they occurred. Illustrations of the devastating effects that measles can have on susceptible young child populations are given in Figure 3.1. Both populations shown here are contemporary but remote, with limited access to healthcare at the times of documented measles epidemics. In the first case, the Wopkaimin population had not been exposed to measles in recent times prior to the establishment of the Ok Tedi gold and copper mine. However, within a year of mining operations starting, infant mortality increased more than seven-fold due to the measles epidemic of 1983. It took five years

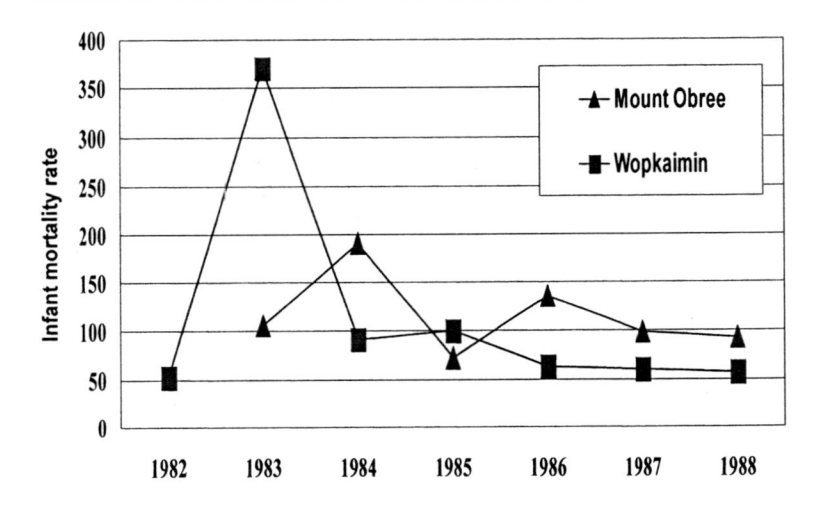

Figure 3.1. *Infant mortality rates among the Wopkaimin and Mount Obree populations, 1982–1988. Adapted from Taufa* et al. *1990*

for the infant mortality rate to decline to the level prior to the epidemic, largely through the implementation of immunisation and primary healthcare, both of which were absent prior to 1983. In the second example, that of the Mount Obree region, the measles epidemic of 1984 saw infant mortality rates rise to nearly double that of the previous year (Taufa *et al.* 1990). Other new diseases against which the Purari population had no resistance included syphilis and tuberculosis. Syphilis, a disease known to reduce female fecundity, became endemic in the Purari delta in the early phases of colonial administration (Hall 1983), while tuberculosis is likely to have been introduced at the time of European contact (Denoon 1989).

The state of food security across this period is not known, although it might be assumed that population decline would have created a surplus of sago palms to harvest, relative to population food needs. However, out-migration of males may have placed an excessive work burden on women, such that food supply may have been limited not by resources, but by the work schedules of women. In 1947, the Nutrition Survey Expedition reported a generally poor nutritional state among the Purari population they visited (Hipsley and Clements 1950), a situation that is likely to have remained unchanged from the start of the twentieth century (Ulijaszek 1993). In such a case, fertility may have been inhibited by poor survivorship of low birth weight infants delivered to undernourished mothers, and to infecundity directly due to the poor nutritional state of women.

Population after the Second World War

From a demographic perspective, the Second World War clearly divides the twentieth century into the period of early colonial administration during which Purari population decline took place, and the subsequent population resurgence during late colonial administration until 1975, and independent nationhood thereafter. Census data for the Purari delta gives a total population estimate of 8,558 for 1920–1924 (Maher 1961), declining to 5,046 in 1955–1956, then rising to 5,776 in 1972–1973 (Hall 1983), and 7,386, from village-based census figures, in 1996. In the late 1940s and 1950s, the combined Baroi and Koriki populations were about half of the combined population of those tribes in 1920–1924 (Maher 1961, Hall 1983). No population values are available between 1920–1924 and 1948–1949, so it is not known whether or not the 1948–1949 and 1951–1952 population sizes represent the twentieth century nadir for the Purari population. However, they may not be far from it. The combined population of the Iari, Kaimari and Maipua tribes underwent a less dramatic decline between 1920–1924 and 1955–1956, but then a more rapid resurgence between 1972–1973 and 1996, bringing the population to around the 1920–1924 value in 1996. The rate of growth of the Purari population was 0.7 percent per year between 1955–1956 and 1972–1973, and 1.2 percent per year between 1972–1973 and 1996. This is very low in contrast to the annual growth rate for Papua New Guinea as a whole; the national rate of population growth was 2.4 percent per year between 1965 and 1980, and 2.2 percent per year between 1980 and 1995 (UNICEF 1997). The low population growth value for the Purari delta masks great variation in rates between the tribes. While the Baroi and Koriki tribes combined grew at 0.3 percent between 1972–1973 and 1996, the combined population of the Iari, Kaimari and Maipua tribes grew by 1.7 percent per year across the same period. Furthermore, the values for overall Purari population growth do not represent true population growth across the second half of the twentieth century, during which time there was a high rate of out-migration.

Differences between rates of mortality and fertility in the Purari delta between 1955–1956 and 1996 indicate a rate of population increase of 1.7 percent per year, a value still lower than the national average, but much closer to it. The difference between known population growth in the Purari delta, and the expected population growth on the basis of differences between crude mortality and fertility values is used here to calculate the excess population that can be presumed to have migrated from the Purari delta. Across the 40 year period, the excess Purari population is estimated to be 2,477, which must have migrated to urban centres, predominantly Port Moresby. In the absence of any indigenous population growth among urban Purari, this suggests that they comprised 23 percent of the total Purari population in 1996, a conservative estimate, since this does not take into account Purari out-migration prior to 1955–1956. Nor does it allow for urban Purari population growth due to excess fertility in relation to mortality.

If it is assumed that the urban Purari population also experienced a 1.7 percent rate of population increase, an additional 704 Purari people would have been born in urban centres (principally Port Moresby), suggesting an urban Purari population of about 3,181, or 30 percent of the total Purari population of 10,567 in 1996. An additional Port Moresby population of about 300 Purari dwellers in Rabia camp in 1955 (Maher 1961), growing again at 1.7 percent per year, is added to this calculation, bringing the total estimated Purari population to 11,129 in 1996, 32 percent of it urban. With increased ease of transport and the ratio of urban to rural Purari people approaching 1 to 2, the urban connectedness of rural Purari has become almost inevitable. Thus, urban migration is very likely to have hidden the true extent of the expansion of the Purari population, which at an estimated 11,000 people, exceeds that of rural Purari in 1920–1924, but remains far from the estimate of around 20,000 given in the Annual Report for Papua, 1907–1908.

The urban Purari population grew after the Second World War because the nature of migration and the mortality risks associated with it were far lower than during early colonial times (Denoon, 1989). To illustrate the extent of out-migration during the second half of the twentieth century, Figures 3.2 and 3.3 show the population pyramids for the Kinipo group of villages of the Purari delta, for the years 1955 and 1996. The extent of male out-migration was enormous in 1955, with 76 percent and 70 percent of the male population aged 10–19 and 20–29 years respectively being absent from the Purari delta (Figure 3.2), most of them living and working in Port Moresby. There was also a significant proportion of out-migrants aged 30–39 years (29 percent of all males). Migration was not confined to males, since 11 percent and 27 percent of females aged 10–19 years and 20–29 years,

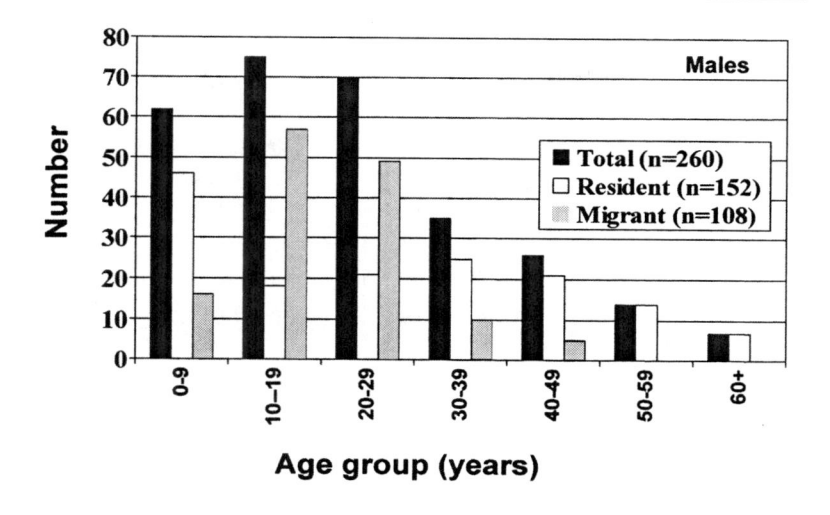

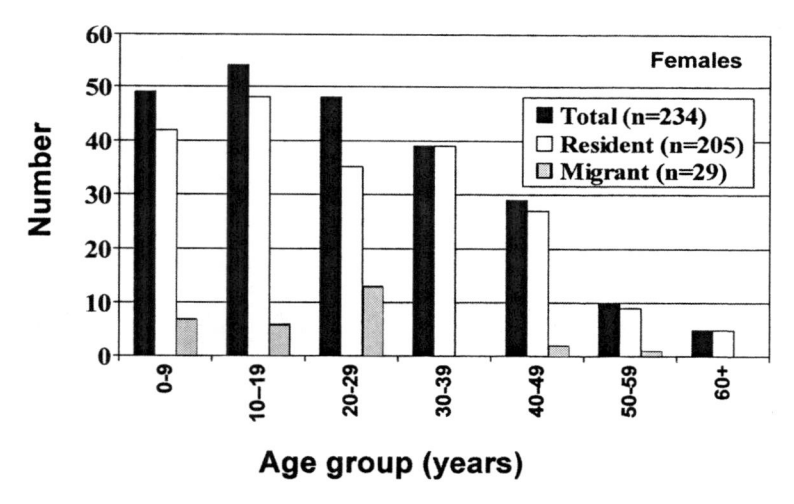

Figure 3.2. *Population pyramid, Kinipo residents, 1955*

respectively, were also absent from the village. Migrants also took children, with a preference for males over females (26 percent males and 14 percent of females aged 0–9 years were also absent from the village). Interestingly, despite such large, overwhelmingly male out-migration, the Purari population began to grow in size. This is probably due to the reduced mortality rates which came with biomedical healthcare when Kapuna hospital was established in 1949.

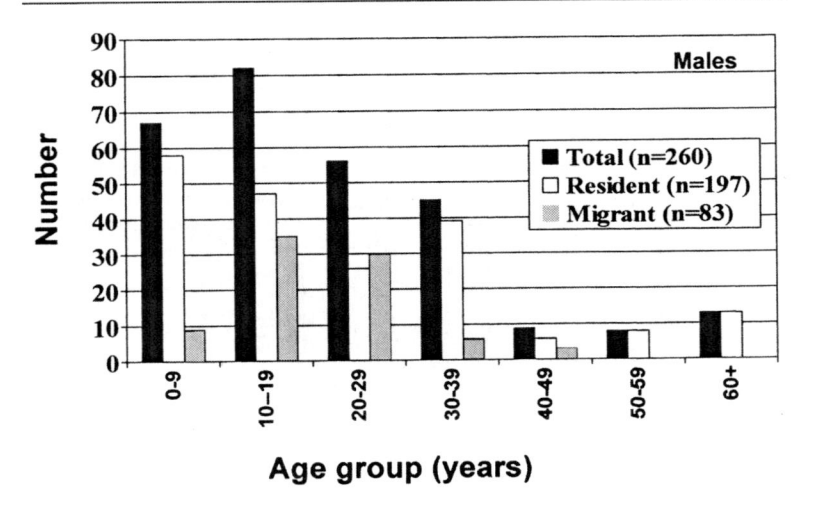

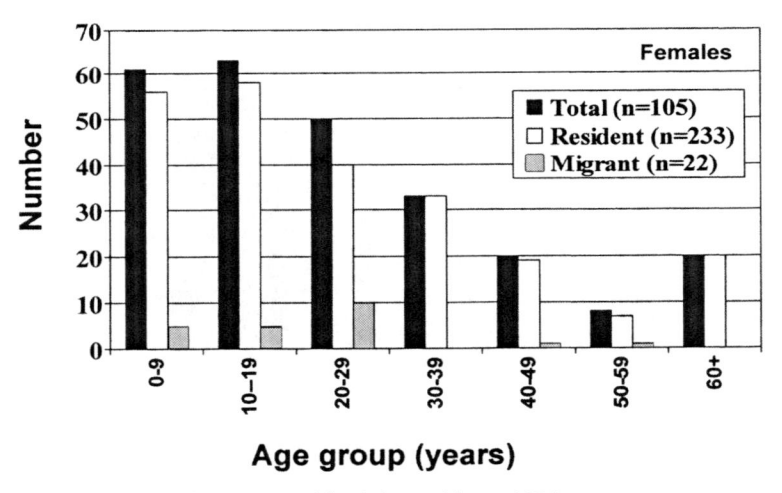

Figure 3.3. *Population pyramid, Kinipo residents, 1996*

Figure 3.3 shows the population pyramids for the Kinipo group of villages in 1996. The extent of young male absenteeism is high, but much lower than in 1955. In 1996, 43 percent of the male population aged 10–19 years had migrated from the Purari delta, while in the age range 20–29 years, 54 percent had done so. There was also significant out-migration of males aged 30–39 years (13 percent), and 40–49 years (33 percent), respectively. As in 1955, migration was not confined to males, with 8 percent and 20 percent of females aged 10–19 years and 20–29 years, respectively, also having migrated. Migrants also took children with them in 1996, with similar proportions of males (7 percent) and females (8 percent) aged 0–9 years having migrated. There were fewer males than females in the Kinipo group of villages in 1996, largely because single males continued to be more likely to migrate for work, or to attend high school. Teenage sons were, and continue to be, more likely to be put forward for high school education by parents, than are daughters. Continued population growth in the presence of high adult male out-migration indicates that reduced mortality, by way of primary healthcare and preventive medicine, continued to be a stronger influence on population size than distorted sex-ratios of marriageable adults keeping fertility rates low.

Evidence for general health improvement across the second half of the twentieth century comes from various sources. The baseline for comparison is the data on health contained in the documentation of the 1947 Nutrition Survey Expedition, which visited one of the Purari villages, Koravake. In the report of this expedition, published in 1950, Hipsley and Clements observed that:

> the general appearance of the people at Koravagi (Koravake) was the reverse of well-being. The majority appeared poorly nourished and underweight. Many, including adults, children and infants, had a severe degree of scabies infestation and impetigo, and many of the children had neglected sores.

Not only did the population appear undernourished at this time, but there was also clear medical evidence of constant exposure to infectious disease. Over half of the Koravake population sampled in 1947 had positive tuberculin tests indicative of TB infection, while 40 percent had palpable spleens, indicative of repeated exposure to malaria (Hipsley 1950). The most frequent species of malaria parasite found in the Purari was *Plasmodium falciparum*, a strong cause of mortality in young children.

The year of the survey, 1947, may have corresponded to the twentieth century nadir in Purari population and possibly health. But it was from this time that the Purari population saw a number of changes, many of them unremarkable in themselves, which changed fertility and mortality rates, and with them, population size and structure. These included the provision of health services, subsistence diversification, and trade. Intensification of palm sago cultivation for consumption and trade, the expansion of sago trade to Port Moresby, as well as the initiation of market food trade to urban centres, including Port Moresby, took place at this time, as did the expansion of small-scale quality fish sales for export. Logging of tropical hardwoods was both initiated and extended, while copra production was expanded. A number of other economic activities were also initiated after 1947, but with limited success, including crocodile farming, and the cash-cropping of rubber, cocoa and chilli peppers.

The major economic event of 1946 was the formation of the Tom Kabu movement, the first significant indigenous attempt to transform the Purari economy (Maher 1961). This attempted, and to some extent succeeded, in stimulating business activities through expansion of existing patterns of sago trade and copra ownership. However, the movement collapsed in 1955, leaving a small number of smaller cooperative organizations running, and patterns of ownership and inheritance remained essentially unchanged. The Tom Kabu movement had, however, stimulated an interest among Purari people in enterprises that were connected to the European economic structure.

Health patterns in the Purari delta began to change after the establishment of a hospital by the London Missionary Society in the Purari delta, in 1949. Prior to this, there had been a permanent establishment to provide biomedical care in the delta for a brief period only, in the form of a temporary hospital in 1921–1922. Medical patrols lead by the colonial administration were infrequent. Kapuna Hospital grew, and began extending community health provision into the surrounding areas. By 1979, five aid posts, serving between 500 and 1500 people at the village level, had been established by Kapuna Hospital across the Purari delta, with an additional aid post having been established by the government. Maternal and child health clinics held by nurses from Kapuna Hospital were started, and have continued to take place on a regular basis in the majority of villages, to the present day. Immunisation and antenatal care of pregnant women became available from 1969 onwards. Collectively, these measures saw a great decline in mortality rate, from a crude death rate of 33 per thousand of population in 1950–1955, to 18 per thousand in 1980–1981 (Hall 1983). Reduced pregnancy loss and increased infant and young child survivorship contributed to the resurgence of Purari population size. There have been improvements in nutritional status (Ulijaszek 1993, 2001) which are likely to have contributed to reduced mortality, by way of increased immunological resistance to infection (Ulijaszek 1990). However, malaria, diarrhoeal and respiratory diseases have continued to be important causes of young child mortality, and collectively continue to help keep population growth below the national level. Malaria and diarrhoeal diseases, including typhoid, continue to be important causes of death in young children.

Factors influencing fertility in the Purari delta in the 1990s

Between the 1950s and 1990s, the population of the Purari delta has seen clear, although sporadic, economic change and increased fertility rates. Crude birth rates have increased from 26 per thousand of population in 1950–1955 (Maher 1961), to 37 per thousand in 1980–1981 (Hall 1983). The TFR was 5.8 in 1966, a value quite typical of simple agricultural societies (Campbell and Wood 1988). It declined very slightly to 5.5 in 1980, where it remained in 1996. This is in contrast to a fertility decline for PNG as a whole, where TFR has fallen from 6.2 in 1966 to 4.7 in 1991. Age-specific fertility rates for women in the Purari delta in 1996 are shown in Figure 3.4. These show Purari women to have their children later than the national average, and much later than both women living in Gulf Province generally, and in urban circumstances in the National Capital District (NCD) (National Statistical Office 1994). The late age-specific fertility of Purari women may reflect the high urban migration rates of males, with consequent late marriage, which is one way in which fertility may have been maintained in early colonial times. Polygyny was another way in which fertility could have been maintained in the face of high male out-migration during colonial times. However, it is unlikely to have continued to be so into recent times. Williams (1924) recorded a polygyny rate of 28 percent among Purari men. By 1955 this had declined to 14 percent of rural Purari men, and only 4 percent of urban Purari males living in Rabia camp (Maher 1961). Estimates of polygyny obtained by questionnaire from males in the Purari villages of Koravake, Mapaio and Ara'ava by the author in 1997 indicate a rate of 5 percent. Thus, polygyny does not appear to have been a response to male out-migration. Rather, the spread of Christianity among the Purari delta populations is likely to have been associated with a decline in polygyny.

Economic change and fertility

While the demographic rebound since the 1950s has exceeded the population estimate for 1920–1924, about a third of the population in the late twentieth century is estimated to be urban, l-

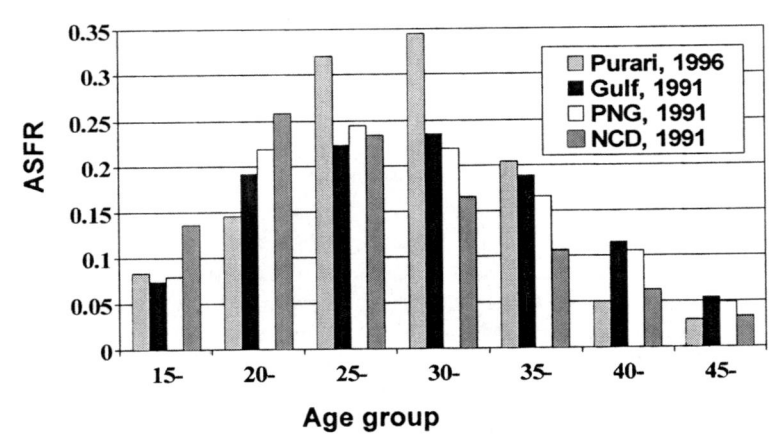

Figure 3.4. *Age-specific fertility rates in Papua New Guinea (Total Fertility Rate: Gulf, Province, 1991 = 4.7; All Papua New Guinea, 1991 = 4.7; National Capital District, 1991 = 4.3; Purari delta, 1996 = 5.5)*

iving predominantly in Port Moresby. While total fertility rates have declined only slightly, economic and social changes have made Purari fertility and population in the 1990s more heterogeneous than in the past. To examine relationships between fertility and such change, statistical analysis of socio-demographic data collected by the author in 1995 and 1997 in the Purari villages Koravake, Ara'ava and Mapaio, was carried out. Multiple regression analysis was performed, using the number of children born as the independent variable, against the following dependent variables: maternal age; number of young children died; income; education; number of urban relatives; and the number of years having lived in an urban centre. The following variables were found to be significantly associated with the number of offspring a woman had: maternal age (standardised beta 0.54, $p < 0.001$); number of young children died (standardised beta 0.17, $p = 0.013$); income (standardised beta 0.14, $p = 0.029$); education (standardised beta -0.15, $p = 0.034$). The full model was statistically highly significant ($r^2 = 0.53$; $F = 38.0$; $p < 0.001$). These statistics show that once the effect of maternal age on number of offspring has been taken into account, the factor most strongly associated with the number of children that women have is child mortality. If a child dies, a Purari woman is very likely to have another child to replace it. While it might be expected that the decline in infant mortality that has taken place with the introduction of healthcare would have reduced the importance of child mortality as an influence on TFR, the multiple regression analysis shows that such a relationship persists in 1995–1997. This indicates that a conflict exists between child death and the need for child replacement, and for smaller desirable family size with increasing maternal education level, despite a large decline in young child mortality rates. The only factor negatively associated with total fertility rate is the educational level of the woman. Figure 3.5 illustrates this effect, showing Purari females who have not completed primary school (with five years or less of school) to start having children earlier and to continue bearing children to a later age than those women who have at least completed their primary school education.

While the TFR is negatively associated with the level of maternal education, the association between income and TFR is not a negative one, as might be expected (multiple regression analysis, and Figure 3.6). The higher the income, the higher the TFR, and the younger the age at which women start to bear children, indicating a tendency for economically successful males to take younger brides. In the Purari, material well-being translates into larger number of offspring, rather than increased investment in a smaller number of offspring, an observation similar to that of Taufa *et al.* (1990) for the Mountain Ok population experiencing rapid economic change in the Ok Tedi Region. However, improved healthcare and increased nutrition have resulted in increased investment in all children of the Purari delta, as demonstrated by the larger body size and improved nutritional status of children of the present generation relative to earlier generations.

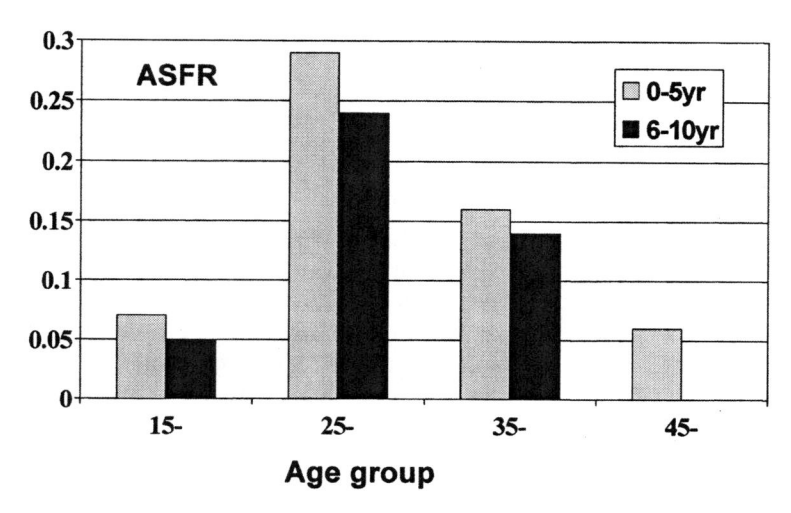

Figure 3.5. *Age-specific fertility and number of years of education, Purari delta. (Total fertility rate: 0–5 years of education: 5.2; 6–10 years of education: 4.3)*

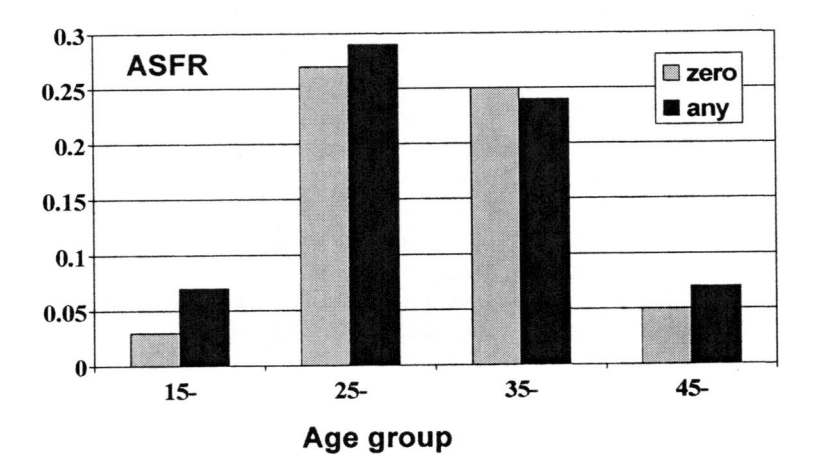

Figure 3.6. *Fertility and income, Purari delta. (Total fertility rate: zero income: 5.5; any income: 5.9)*

Urban connectedness, through relatives, or migration at some time of life, seems to have no influence on TFR, despite the expectation that exposure to outside ideas concerning appropriate family size might be a force for reducing TFR, as public health rhetoric might have it. Another reason for the maintenance of high TFR among the Purari, unexplored by fieldwork, may be the breakdown in ritual prohibitions in the past 50 years or so. While the Ravi existed in the fortressed villages prior to European pacification, both men and women observed taboos concerning sexual conduct and interaction. With the abandonment of these structures and the formation of new types of settlement, these restrictions have faded.

Subsistence and fertility

Palm sago is at the centre of Purari subsistence ecology. Existing patterns of sago tenure and use link population to resources in a very direct way, in that sago-gardens may be inherited at marriage and new sago gardens are often planted when children are born, at least among the Baroi tribe. In the Purari delta, increased land has been devoted to sago cultivation since 1947, and possibly earlier. The density of planting and number of palm-sago cultivars used has also changed with time. Sago gardens planted between 1978 and 1996 contain a maximum of three cultivars, from 12 possible cultivars known to have been planted in gardens created prior to then. These three cultivars are seen to be higher yielding and 'easier to work' than most of the ones that are no longer cultivated (Ulijaszek 1991). This intensification of sago-palm use has accompanied increased population size, at least since 1947. The fact that women carry the majority of the burden of sago-making (Ulijaszek and Poraituk 1993) may link increasing food demands, which are largely met by sago production, to female reproductive physiology. The traditional pattern of women's work in sago production, fishing and other foraging accommodates a pattern of frequent and prolonged breastfeeding, which reduces ovulatory function and the likelihood of conceiving (Ellison 2003). While women have continued to be engaged in subsistence activity, the opportunity for extended lactation remains, as does the ovulatory inhibition that accompanies it. The heavy workload associated with sago-making by women (Ulijaszek 1997) is a further factor that may inhibit ovulation. However, counter to this is the improved nutritional condition of the women since about 1980, and the decline in breastfeeding duration that has taken place among more educated women, both factors serving to enhance ovulatory function and increase the potential for conception. More men than before engage themselves in fishing, and this may also contribute to improved nutritional status of women.

In the Purari delta, palm sago has a dual role as both staple and commodity, and there is no diversion of female labour from the usually flexible physical demands of sago-making to another productive activity, as for example with coffee cash-cropping in the New Guinea Highlands (Johnson 1990). Although women may be processing sago on a more regular basis in recent times than in the known past, sago-making as income generation does not interfere with the need to breastfeed on demand. Furthermore, sago-making is not very seasonal, being limited only by a lack of clean water for its processing for a limited period in the drier season. There is variation in the time spent and intensity of work performed by Purari women in sago-making (Ulijaszek and Poraituk 1993), harder-working women generating more dietary and nutritional diversity by fishing and horticulture on the days not spent sago-making than women who work less hard (Ulijaszek 2003a). Although this is not reflected in the anthropometric nutritional status of their children, such diversity is needed to give nutritional balance to a diet that would otherwise be limited in all nutrients apart from energy (Ulijaszek 1983), and may confer greater immunological protection from infection among all family members (Ulijaszek 1990). Purari women also perform sago-making as households, sisters, mothers, daughters and sisters-in-law working cooperatively to speed up the work. Women (and female members of households) that spend longer in processing sago starch on any particular day are freed up from making sago on other days, because they can generate more staple food on that day. This may give them more time to generate income by other means, such as marketing foods in a number of places. The longest standing of these are at the nearest urban centre, Baimuru, and at Kapuna. Between the mid-1990s to the year 2000, Purari people travelled to Kikori to take advantage of market opportunities in that area resulting from both the oil pipeline developed, and the increased logging activity in that district. From the year 2000 to the time of writing, people from Mapaio, Kinipo and Maipenairu have been travelling to the Frontier Holdings base camp on the Purari River for similar opportunities.

Greater income from market- and cash-cropping may well translate into greater nutritional availability, probably in the form of bought trade-store foods such as rice, hard biscuits, tinned fish and tinned meat, and may be associated in this way with greater weight and body mass index of adults (Ulijaszek 2003b). Better-nourished women engaged in a flexible mix of subsistence and income-generating production in the Purari delta are thus likely to be able to continue prolonged breastfeeding, allowing greater survivorship of their offspring. Prolonged lactation and improved maternal nutritional status have effects on ovulatory function which oppose each other, the former being inhibitory, the latter promoting it. Thus, economic development in the Purari delta appears to be associated with sustained maternal fertility levels combined with improved child survivorship.

Concluding remarks

The population decline in the early colonial period in the Purari delta can largely be attributed to infectious diseases introduced with the contact experience, and to low crude birth rates associated with the recruitment of adult males for plantation labour. This is in accord with a 'fatal impact' perspective of European colonisation of the Pacific (Moorehead, 1966). However, as elsewhere in Melanesia (see, for example, Gosden, this volume), people in the Purari delta reconfigured themselves physically and economically, and with respect to village structure and location, engagement in cash-cropping and trade, as well as in migration to

Port Moresby and elsewhere in PNG. These factors have had indirect inputs to population growth in the second half of the twentieth century, by contributing to improved nutrition, which is likely to have increased the fecundity of women, and increased resistance to diseases carrying high mortality rates among children. The introduction and increased availability of biomedical practice has also had a major influence on population growth, by way of increased survivorship of young children. Despite a decline in mortality, there has been no accompanying decline in fertility rates in the second half of the twentieth century. Young child mortality and income availability are important factors supporting the maintenance of high fertility in the 1990s, while maternal education is associated with reduced TFR. Thus there is a conflict between child death and the desire among the Purari for child replacement, and for smaller desirable family size with increasing education level of the mother.

Acknowledgement

I am grateful to Josh Bell for discussion of this paper, and for various useful suggestions made by him for its improvement.

References

Adels, B.R. and D.C. Gajdusek 1963. 'Survey of Measles Patterns in New Guinea, Micronesia and Australia', *American Journal of Hygiene* 77: 317–343.

Annual Report for Papua 1907–1908. Melbourne: Government Press of Victoria.

Ashford, R.W. and D. Babona 1980. *Viral and Parasitic Infections of the People of the Purari River and Mosquito Vectors in the Area*. Purari River (Wabo) Hydroelectric Scheme Environmental Studies, Volume 8. Waigani, Papua New Guinea: Office of Environment and Conservation.

Campbell, K.L. and J.W. Wood 1988. 'Fertility in Traditional Societies'. In *Natural Human Fertility. Social and Biological Determinants*, eds P. Diggory, M. Potts and S. Teper, 39–69. London: The Macmillan Press.

Denoon, D. 1989. *Public Health in Papua New Guinea. Medical Possibility and Social Constraint, 1884–1984*. Cambridge: Cambridge University Press.

Durrad, W.J. 1922. 'The Depopulation of Melanesia'. In *Essays on the Depopulation of Melanesia*, ed. W.H.R. Rivers, 3–24. Cambridge: Cambridge University Press.

Ellison, P.T. 2003. 'Energetics and Reproductive Effort', *American Journal of Human Biology* 15: 342–351.

Hall, A.J. 1983. 'Health and Diseases of the People of the Upper and Lower Purari'. In *The Purari – Tropical Environment of a High Rainfall River Basin*, ed. T. Petr, 493–507. The Hague: Dr. W. Junk Publications.

Hipsley, E.H. 1950. 'Report on Health and Nutritional Status in New Guinea'. In *Report of the New Guinea Nutrition Survey Expedition, 1947*, ed. E.H. Hipsley and F.W. Clements, Part 5, 143–176. Sydney: Government Printer.

———. and F.W. Clements 1950. *Report of the New Guinea Nutrition Survey Expedition, 1947*. Sydney: Government Printer.

Hope, G.S., J. Golson and J. Allen 1983. 'Palaeoecology and Prehistory in New Guinea', *Journal of Human Evolution* 12: 37–60.

Johnson, P.L. 1990. 'Changing Household Composition, Labor Patterns, and Fertility in a Highland New Guinea Population', *Human Ecology* 18: 403–416.

Keeling, M.J. and B.T. Grenfell 1997. 'Disease Extinction and Community Size: Modelling the Persistence of Measles', *Science* 275: 65–67.

Knauft, B.M. 1993. *South Coast New Guinea Cultures*. Cambridge: Cambridge University Press.

Maher, R.F. 1961. *New Men of Papua*. Madison: University of Wisconsin Press.

Moorehead, A. 1966. *The Fatal Impact: an Account of the Invasion of the South Pacific, 1767–1840*. London: Hamilton Press.

National Statistical Office 1994. *Report on the National Population and Housing Census in Gulf Province*. Port Moresby: National Statistical Office.

Oram, N. 1992. 'Tommy Kabu. What Kind of Movement?', *Canberra Anthropology* 15: 89–105.

Rhoads, J. 1982. 'Sago Palm Management in Melanesia: an Alternative Perspective', *Archaeology in Oceania* 17: 20–27.

Riley, I.A., D. Lehmann and M.P. Alpers 1992. 'Acute Respiratory Infections'. In *Human Biology in Papua New Guinea. The Small Cosmos*, eds R.D. Attenborough and M.P. Alpers, 281–288. Oxford: Oxford University Press.

Rivers, W.H.R. 1922. 'The Psychological Factor'. In *Essays on the Depopulation of Melanesia*, ed. W.H.R. Rivers, 84–113. Cambridge: Cambridge University Press.

Taufa, T., V. Mea and J. Lourie 1990. 'A Preliminary Report on Fertility and Socio-economic Changes in Two Papua New Guinea Communities'. In *Fertility and Resources*, eds J. Landers and V. Reynolds, 35–46. Cambridge: Cambridge University Press.

Ulijaszek, S.J. 1983. 'Palm Sago (Metroxylon Species) as a Subsistence Crop', *Journal of Plant Foods* 5: 115–134.

———. 1990. 'Nutritional Status and Susceptibility to Infectious Disease'. In *Diet and Disease*, eds G.A. Harrison and J.C. Waterlow, 137–154. Cambridge: Cambridge University Press.

———. 1991. 'Traditional Methods of Sago Palm Management in the Purari Delta of Papua New Guinea'. In *Proceedings of the Fourth International Sago Symposium, Kuching, Sarawak, Malaysia*, eds N. Thai-Tsiung, T. Yiu-Liong and K. Hong-Siong, 122–126. Sarawak: Ministry of Agriculture and Community Development, and Department of Agriculture.

———. 1993. 'Evidence for a Secular Trend in Heights and Weights of Adults in Papua New Guinea', *Annals of Human Biology* 20: 349–355.

———. 1997. 'Energy Expenditure in Sago Processing by Women in the Purari Delta of Papua New Guinea', *Anthropological Science* 105: 143–147.

———. 2001. 'Secular Trend in Birthweight among the Purari Delta Population, Papua New Guinea', *Annals of Human Biology* 28: 246–255.

———. 2003a. 'Maternal Work and Childhood Nutritional Status among the Purari, Papua New Guinea', *American Journal of Human Biology* 15: 472–478.

———. 2003b. 'Socioeconomic Factors Associated with Physique of Adults of the Purari Delta of the Gulf Province, Papua New Guinea', *Annals of Human Biology* 30: 316–328.

——— and S.P. Poraituk 1983. 'Subsistence Patterns and Sago Cultivation in the Purari Delta'. In *The Purari – Tropical Environment of a High Rainfall River Basin*, ed. T. Petr, 577–588. The Hague: Dr. W. Junk Publications.

——— and S.P. Poraituk 1993. 'Making Sago: Is It Worth the Effort?' In *Tropical Forests, People and Food. Biocultural Interactions and Applications to Development*, eds C.M. Hladik, A. Hladik, O.F. Linares, H. Pagezy, A. Semple and M. Hadley, 271–280. Paris: UNESCO Publications.

UNICEF. 1997. *The State of the World's Children 1997*. Oxford: Oxford University Press.

Williams, F.E. 1924. *The Natives of the Purari Delta*. Anthropological Report No. 5. Port Moresby: The Government Printer.

CHAPTER 4

MIGRATION AND FERTILITY OF A SMALL ISLAND POPULATION IN MANUS: A LONG-TERM ANALYSIS OF ITS SEDENTES AND MIGRANTS

Yuji Ataka and Ryutaro Ohtsuka

The second half of the twentieth century has seen rapid population increase, with associated migration from rural to urban areas, in Papua New Guinea (PNG). The population growth rate in the urban sector accelerated especially in the 1960s (Skeldon 1982a), with the majority of rural to urban migrants settling in the national and provincial capitals (Skeldon 1982b; Walsh 1987a, 1987b), resulting in 15.4 percent of the national population living in urban places in 1990. The majority of migrants to urban centres remain for several years or longer, there being negligible seasonal migration. Strong social relations between urban migrants and their rural counterparts are largely maintained according to the *wantok* system that operates within a language group. Physical manifestations of such relations include the payment of remittances by urban residents to their rural relatives (Carrier and Carrier 1989), and the great popularity among migrants of returning to their rural homeland for Christmas and the New Year, and for ceremonial occasions including rites of passage, marriages, funerals, and ancestral worship. These opportunities ensure the transmission of information about births, deaths, and marriages of *wantok* members, as well as details of daily living conditions in both locations.

A number of demographic studies in developing countries have generally shown lower levels of fertility in urban than in rural areas (e.g. Brockerhoff and Yang 1994; Goldstein and Goldstein 1981; Lee and Farber 1984). However, there have been few studies in which fertility of both migrants and non-migrants of the same population or community have been compared (Ohtsuka 1994a). In one such study, among the Kombio of East Sepik Province of PNG, Umezaki and Ohtsuka (1998) found lower fertility in migrants than in sedentes, supporting the view that rural to urban migration tends to reduce the overall rate of increase of the rural and urban populations combined (Bogin 1988).

In this chapter we describe the changing vital events of sedentes and migrants to urban areas, of an island population living in Perelik village, on Baluan Island of Manus Province (previously Manus District), across the period 1955 to 1995. The interrelationships between rural population density, the changing nature of the migrant urban population and fertility are explored.

Manus: Geographical and historical backgrounds

Location and people

Manus Province is situated just south of the equator, covering the area of 141°–149° E longitude and 0°–4° S latitude, in the northernmost part of PNG (Figure 4.1). This province comprises Manus Island and a large number of small coral and volcanic islands, more than 200 in number, scattered over 220,000 square kilometres of the sea area despite the fact that the total land area is only 2,100 square kilometres. According to meteorological records at Momote (at the eastern end of Los Negros Island), annual rainfall reaches 3,300 mm without marked fluctuation across the year, while mean maximum and minimum monthly temperatures are 29.9°C and 24.4°C, respectively (Manus Provincial Government 1982). The Manus Islands populations are broadly classified into three (Mead 1934; Mitton 1979): the Manus, who have traditionally subsisted on fishing and trading and have inhabited the southern coast of Manus Island and some offshore islands; the Usiai, who have specialised in horticulture in the interior of Manus Island; and the Matankor, who have conducted both slash-and-burn horticulture and fishing, living on small islands which are north, east, and southeast of Manus Island. The last forms the study population. Until the recent past, each group

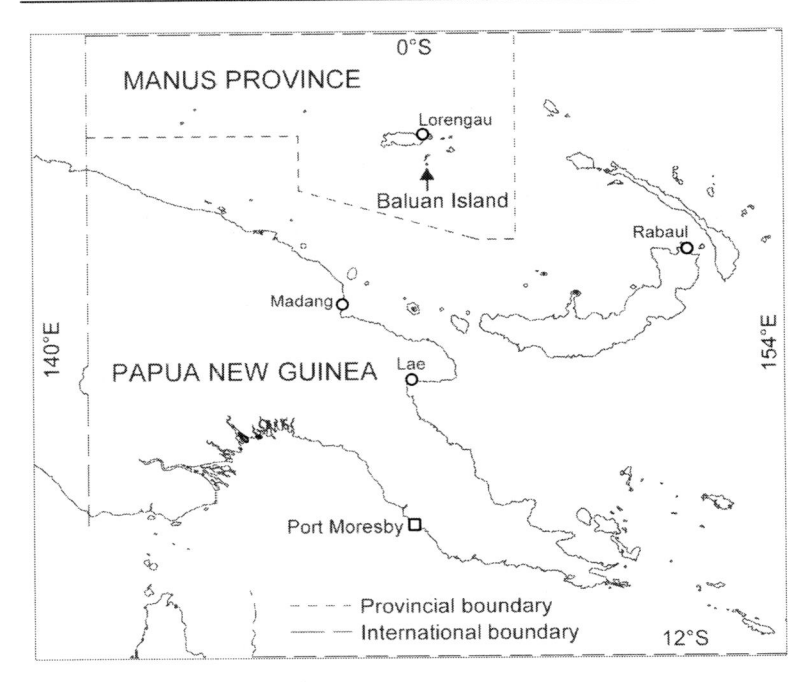

Figure 4.1. *Location of the Baluan Island and Manus Province, Papua New Guinea*

relied on the others by means of trade and exchange of foods such as marine products, horticultural crops, and *Metroxylon* sago flour, as well as exchange of various local non-food products.

Contact history

The Manus Islands were first sighted by Europeans, led by W.C. Schouten, a Dutch navigator, in 1616, although it has also been suggested that Á. Saavedra, a Spanish explorer, sighted them in 1528. However, the frequent contact of these islanders with Europeans only began in the nineteenth century, with European traders seeking marine resources such as trepang (bêches-de-mer) and tortoise shells, as well as coconuts for making copra. The Islands region of New Guinea, including Manus, were governed as a German protectorate from 1884, German traders and merchants establishing a number of commercial ventures, primarily coconut plantations (Firth 1983). In 1914, the supreme power of these islands was ceded to Australia, and they became part of the territory mandated to that nation in 1921. By the Second World War, a considerable number of Manus Islanders, mostly young males, had already migrated, or experienced migration, to other areas of New Guinea as contract labourers in plantations and to the so-called 'European towns' in Rabaul and other areas, as housekeeping servants. In 1946, the Manus Islands became a part of the United Nations Trust Territory of New Guinea under the control of Australia. Since 1975, when PNG attained its independence, all of the Manus Islands formed one province of this new nation.

One of the most influential European acculturation impacts was the introduction of education, along with Christianity, in the early twentieth century (Carrier and Carrier 1989). On Baluan Island, where the study population resides, primary education was established in 1951. Immediately after the Second World War, the Paliau movement was started in Manus. This started as a radical reformation campaign aimed at emancipating the Manus people from colonial inequality, which then became an important political, economic, and, in some aspects, religious movement (see Schwartz 1962 for details). The high level of agency expressed by people in the Islands region is reflected in their current living standard and health status. In the 1990s, the enrolment and completion rates for primary school education in Manus were the highest of all the provinces in the Islands region, although the literacy rate was highest in East New Britain. According to health statistics for PNG for the year 1990, the infant mortality rate was lower and life expectancy at birth was greater in the Islands region than in any other region of PNG (Table 4.1). Among the provinces of the Islands region, Manus had the low-

Table 4.1. *Infant mortality rate, life expectancy at birth, and annual increase rate, broken down by the four regions of Papua New Guinea*

Region	Infant mortality rate* (1990)	Life expectancy at birth* (1990)		Annual increase rate from 1980 to 1990		
		Males	Females	Rural	Urban	Total
Papua†	51–111	43.0–60.5	46.5–65.6	1.8	4.9	2.7
Highlands	75–110	42.1–52.0	45.4–58.3	2.0	2.3	2.0
Momase	81–99	46.1–50.0	47.5–59.4	1.5	3.4	1.8
Islands	40–70	53.9–65.9	55.8–62.9	3.0	5.0	3.3
Total‡	82	52.3	53.9	2.0	4.0	2.2

Note: Continuous annual increase rate.
* According to the Health and Demographic Survey 1991. The islands region excluded the North Solomons Province since there was no data due to the Bougainville crisis.
† Excluded the National Capital District (NCD), Port Moresby.
‡ NCD was included, but North Solomons Province was not included.

est infant mortality rate (40 per thousand live births, for both sexes combined) and the greatest life expectancy at birth (65.9 years for males and 62.9 years for females); the latter levels were almost identical to those for Port Moresby. The rate of population increase was the highest in the Islands region, compared with all other regions of PNG.

Rural to urban migration

Of all provinces of PNG, Manus is characterised by the smallest land area and the remotest location from the main island of New Guinea. There are few opportunities for obtaining work requiring high levels of training or skill within the province, especially among young people, and despite the good levels of education attained by many of them. It is unsurprising, therefore, that there is a long history of Manus people migrating to remote urban centres for employment, as well as for higher levels of education, even prior to the nation's independence in 1975 (Carrier and Carrier 1989; National Statistical Office 1994; Skeldon 1982b). Major destinations for out-migrants past and present were, and continue to be, Port Moresby and provincial capitals such as Rabaul (East New Britain), Madang (Madang), and Lae (Morobe). The majority of migrants either attend high school or university, or work as public servants, teachers or self-employed businessmen.

The setting

Perelik: the location and population

Baluan is a volcanic island, located 40 kilometres south of Manus Island (Figure 4.1); it takes three to five hours to travel from Baluan to Lorengau, the capital and commercial centre of Manus Province, by small motor boat. There are seven villages on the island. Of these, six have been inhabited by Matankor people and one by Manus people. Perelik village was selected from the six Matankor villages for this study. Perelik villagers, as many other Matankor people in Baluan, have subsisted on slash-and-burn cultivation of tuberous crops and bananas, and fishing. The sale of cash crops such as copra and cacao to middlemen and the sale of food crops at local markets on the island and in Lorengau have played supplementary roles in their economy. Trade of marine resources, especially trochus shells, has gradually expanded across the last two to three decades (Ataka and Ohtsuka 2000).

The Perelik people have been, in greatest part, believers of the Seventh-Day Adventist (SDA) faith since the late 1930s. This

Christian sect prohibits its followers to eat pigs, crustaceans, shellfish, and some types of fish. It also encourages people to work hard and to modernise their lifestyles. Because of the modernising influence of the SDA church, many villagers know their date or year of birth, making the collection of demographic data more accurate here than among many PNG populations.

Methods

The field research was conducted mostly on Baluan Island but occasionally in other areas of Manus Province, particularly Lorengau, where many people of Perelik origin reside. A demographic survey was carried out across 24 weeks during 1994 and 1995, with an additional short-term survey, which was undertaken in 1997. During fieldwork, the villagers' subsistence patterns and health problems, mostly related to malaria prevalence, were also investigated (Ataka and Ohtsuka 2000; Ataka et al. 2001).

Demographic data collection involved the reconstruction of detailed genealogies with time-depth (Ohtsuka 1986; Umezaki and Ohtsuka 1998). Firstly, an extensive interview survey was conducted to identify all lineages and sub-lineages of Perelik origin, and the founders of these lineages. As far as was possible, past members of each lineage were then identified. It was possible to identify all current Perelik villagers with their lineages. A detailed interview survey of each adult Perelik villager was subsequently carried out to clarify their genealogical relations and to record the dates (in years) of any vital events of themselves, and their kin, including relatives who had died and those who had migrated from Perelik. In the latter survey, the elderly villagers in particular provided useful information. Furthermore, information about vital events of migrants was collected when they temporarily returned to Baluan for ceremonial occasions or for Christmas.

The majority of villagers that were interviewed remembered or recorded the dates (in years) of their own and their relatives' vital events (especially births and deaths) since the 1930s, when the SDA church was established. For questionable dates of events, most of which took place long ago, estimates were made using an 'event history', constructed by the authors, from key events given in government patrol reports and the years of well-known local events such as a big feast held in 1946 and the establishment of the first primary school in Baluan in 1951. The estimation of the year of a vital event was then repeatedly conducted until the three following kinds of information became consistent: the birth order of all individuals listed in the genealogical charts; the birth years of all of them, estimated from the event history; and the years of their death, marriage, divorce, and adoption, if any, estimated in the same way.

Similarly, the migration histories of all individuals listed in the genealogical charts were reconstructed by interview with current village members, other persons who inhabited the neighbouring islands and Lorengau, and migrants to remote areas who temporarily returned to Perelik. Out-migration was defined as the movement of residence out of Baluan Island for six months or longer. The residential locations were identified, in detail, with local place names. For the purposes of this analysis, members of the Perelik population were defined as persons who had landowning rights in the village territory. According to the tradition of their patrilocal lineage system, only married male members of the landowning lineages have such rights. Thus, the study focused on such males and their spouses and offspring. Females married to the spouses belonging to the Perelik landowning lineages were included, while females of the Perelik landowning lineages were excluded if they were married to the spouses who did not belong to these lineages. For subsequent analysis, residential locations were classified into the rural sector (Baluan Island), and the urban sector which included all destinations of migrants. There were a few migrants, mostly teachers and public servants, who lived in non-urban areas, but it was decided to place them in the urban sector for analysis, because their type of work and living conditions more closely resembled the urban than rural situation.

The demographic data analysed in this paper were vital events and migrations (out-migrations and return-migrations) of all Perelik members from 1955 to 1994, comprising a sample of 355 people. Due to the small number, the time unit of analysis was set at a 10-year interval: 1955–1964, 1965–1974, 1975–1984 and 1985–1994. As an example, the crude birth or death rate in 1955–1964 was obtained by dividing the sum of birth or death cases by the person-years of all individuals for the 10-year duration concerned.

Population increase

Figure 4.2 shows changes of the Perelik population from 1955 to 1995 according to location. The population as a whole was 80 in 1955 and 210 in 1995. Population increase was much higher in the urban sector than in the rural sector, especially after 1970. The rural-sector population increased rapidly between 1985 and 1994, even though the rate of increase was lower compared to that in the urban sector across the same period (Table 4.2). The

Migration and Fertility of a Small Island Population 97

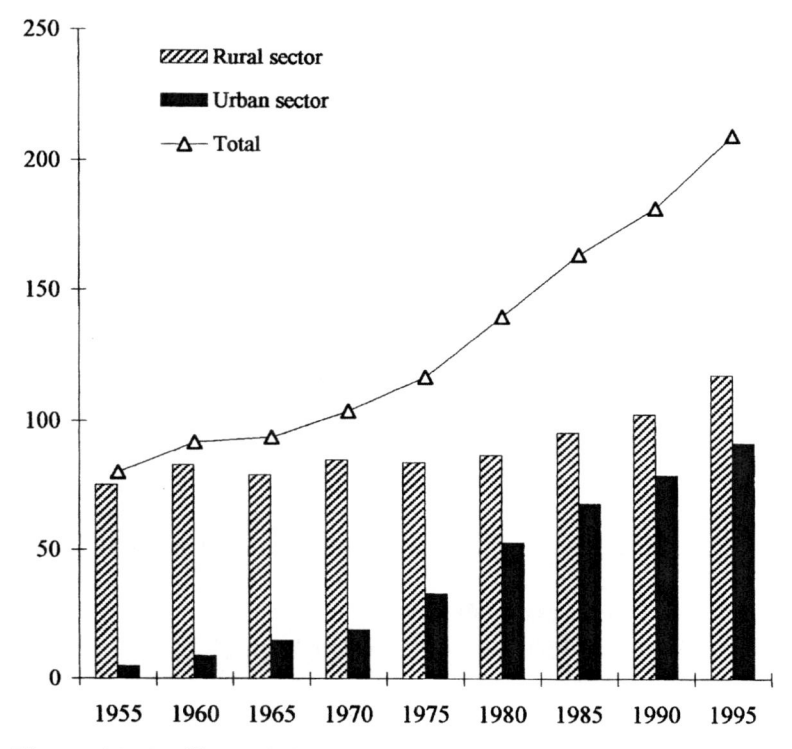

Figure 4.2. *Perelik population size, 1955–1995*

proportion of urban-sector dwellers to the whole Perelik population increased from only 6 percent (5/80) in 1955 to 44 percent (92/210) in 1995. The urban Perelik population grew by 7.3 percent between 1955 and 1994, while the rural population grew by only 1.1 percent. The mean annual increase rate of the whole Perelik population was 2.4 percent over the 40 years from 1955 to 1994, the corresponding population doubling time being approximately 29 years. The population increase rate for the 10-year

Table 4.2. *Annual population increase rate in the rural and urban sectors for four 10-year periods and the whole duration, 1955–1994*

Residential sector	1955–1964	1965–1974	1975–1984	1985–1994	1955–1994
Rural	0.5	0.6	1.3	2.1	1.1
Urban	11.0	7.9	7.2	3.0	7.3
Both	1.6	2.2	3.4	2.5	2.4

Note: Continuous annual increase rate

period of 1985–1994 was 2.5 percent, being almost identical with those of Manus Province (2.3 percent) and PNG as a whole (2.2 percent) in the 10-year period between 1980 and 1990.

The balance of the number of women entering and leaving the Perelik population by marriage, including some child adoption, accounted for only 10 percent (13/130) of the population increase across the whole observation period. The positive balance between 1955 and 1994 is the sum of out-migration in the two earlier decades (−7 in 1955–1964 and −3 in 1965–1974), and in-migration in the two later decades (+9 in 1975–1984 and +14 in 1985–1994).

Birth rate and death rate

The crude birth rate (per 1000) of the rural and urban Perelik population combined was high in the three earlier periods: 35.3 in 1955–1964, 32.1 in 1965–1974, and 35.4 in 1975–1984. This decreased to 29.1 in the 1985–1994 period. The crude death rate (per 1000) decreased from 13.3 in 1955–1964 to 6.6 in 1975–1984, slightly increasing to 8.2 in 1985–1994. The marked decrease in mortality between the earlier two decades to the later two decades was due mostly to reduced young child mortality. The mean under-five years mortality rate of the Perelik population in 1955–1974 and 1975–1994 was 123 and 32 per 1,000 live births, respectively. The under-five years mortality rates in 1971 and 1980 were 56 and 32, respectively, in Manus, and 79 and 42, respectively, in PNG as a whole (Ministry of Health 1994). A sharp decline in child mortality in Perelik might be attributed to a decrease in deaths due to malaria. Malaria control by residual spraying began in Manus in the 1960s, following a pioneering project of this type in the Maprik area of East Sepik District in 1957. Residual spraying was successful in the earliest stages on the Manus Islands, leading to a dramatic reduction in the malaria parasite rate, from 16.7 percent in 1961 to 2.9 percent in 1972–1973 (Parkinson 1974). However, Manus continues to be a malaria endemic area (Ataka et al. 2001; Manus Provincial Government 1982), and the low child mortality rate in the period 1975–1994 may be primarily due to improved health services, including the availability of antimalarial drugs. The reduced rate of Perelik population increase in the period 1985–94 relative to the period 1975–1984 was due largely to decreased fertility, a consequence of the successful uptake of family planning practices.

Changes in migration pattern

The number of rural to urban migrants increased rapidly from around 1975, although their destinations varied across time. The proportion of migrants who left the Manus Islands was 20 percent (1/5) in 1955, 60 percent (9/15) in 1965, 48 percent (16/33) in 1975, 85 percent (58/68) in 1985, and 58 percent (53/92) in 1995. Although the migration of Perelik villagers to take up contract labour in plantations or to be housekeeping servants in the European towns had ended by 1955, the year from which data for the present analysis was collected (cf. Ward 1977), contract labour migrations to plantations continued until recent times in PNG more generally (e.g. Curry and Koczberski 1998).

Figure 4.3 illustrates the increasing number of urban-sector dwellers across the period 1955 and 1995, according to purpose of migration, or occupational status in the urban setting. Salaried workers and their families accounted for two-thirds or more of all migrants, followed by students and then by migrants without specific purpose (who were jobless or, at least, did not have regular occupation). Taking into account the number of rural to urban

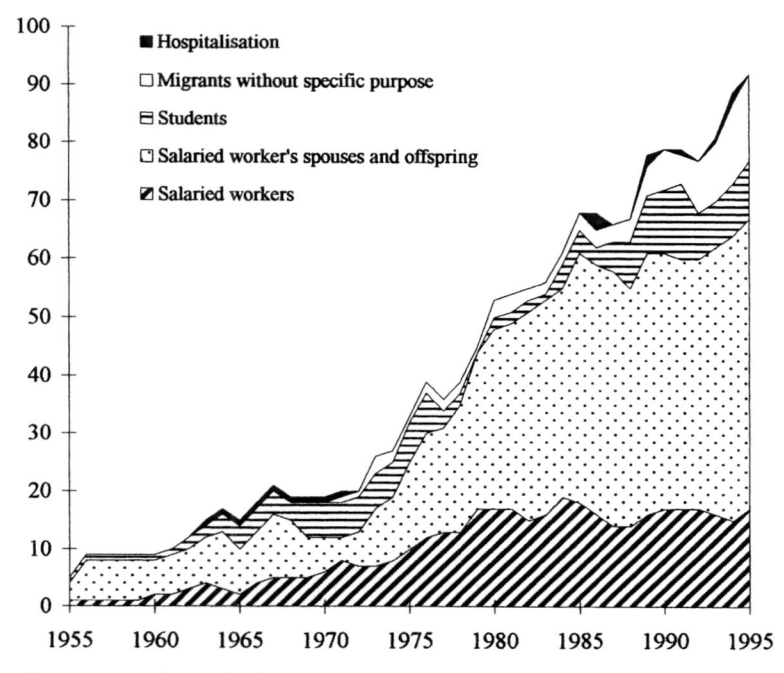

Figure 4.3. *Change in the number of persons in the urban sector from 1955 to 1995, according to occupational status*

migrants (or urban dwellers) and their breakdown by occupational status, the 40-year observation period was divided into three phases: from 1955 to 1975, from 1975 to 1985, and from 1985 to 1995.

The first phase was characterised by a small but gradually increasing number of migrants. The proportion of students, most of whom were males and high school and university students, was relatively high, at 20 percent or more, partly because the opportunities of receiving higher education were enlarged; for instance, the University of Papua New Guinea was opened in Port Moresby in 1965. The second phase was characterised by an increase in salaried workers and their families. This increase can be attributed in largest part to changes in the migrants' status from students to salaried workers, and to a much lower extent to direct migration from Perelik. The urban Perelik population was boosted by spouses and offspring of salaried workers to a much higher extent than by additional salaried workers themselves. Although the rate of creation of employment opportunities in urban areas (apart from Port Moresby and Kavieng) may have slowed down since the early 1970s (Garnaut et al. 1977), the number of salaried workers from Perelik increased in this phase, with little fluctuation.

In the third phase, the changes in the proportions of Perelik-origin urban dwellers by job type or purpose of migration differed from those in the second phase. Although salaried workers and their spouses and offspring were unchanged in number, students and migrants without jobs increased. The proportion of the latter two groups combined accounted for more than a quarter of all Perelik migrants in 1995. The major reason why the number of salaried workers did not increase and the number of jobless migrants did increase is that population pressure in Perelik village pushed an increasing number of new migrants to compete for a very limited number of new formal-sector jobs in urban areas at this time. Despite significant localisation of many public servant positions after the expatriate exodus post-independence in 1975, the rate of increase of formal-sector employment opportunities has been slow and has gradually dropped due to economic stagnation in recent years. There were several salaried workers who returned to Perelik in the latter half of the 1980s, when the Bougainville crisis took place.

Differential fertility between urban and rural Perelik

Figure 4.4 shows age-specific fertility rates (ASFR) for rural and urban Perelik people for the four 10-year periods. The ASFR of urban-sector women are excluded in cases where the total

Migration and Fertility of a Small Island Population 101

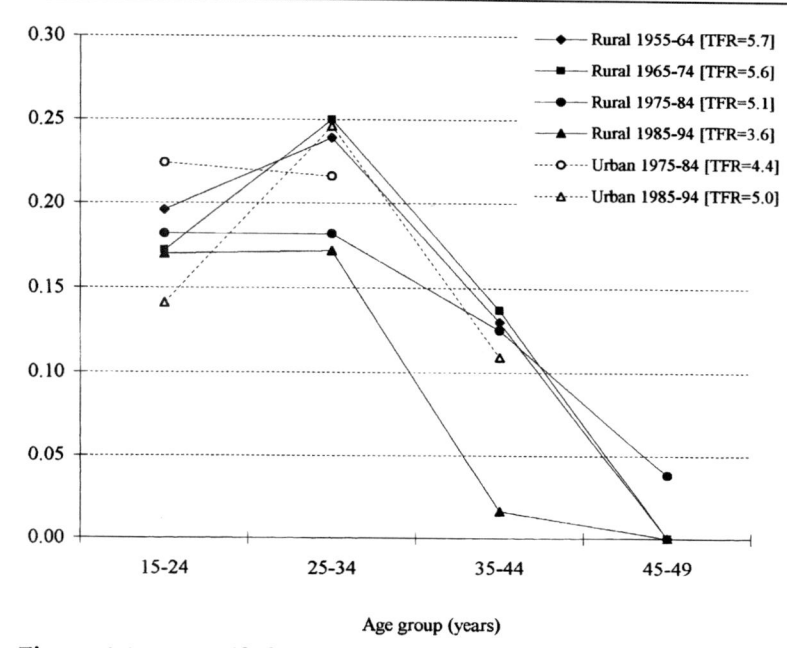

```
                    ◆ Rural 1955-64 [TFR=5.7]
                    ■ Rural 1965-74 [TFR=5.6]
                    ● Rural 1975-84 [TFR=5.1]
                    ▲ Rural 1985-94 [TFR=3.6]
                    ○ Urban 1975-84 [TFR=4.4]
                    △ Urban 1985-94 [TFR=5.0]
```

Age group (years)

Figure 4.4. *Age-specific fertility rate (ASFR) of women in the rural and urban sectors in the four 10-year periods. The rates for the urban-sector women are not mentioned in cases that the woman-years were less than 40. Total fertility rate (TFR) was calculated from the ASFR values from the four age groups for the rural sector but from those of 15–24 and 25–34 age groups and 15–24, 25–34, and 35–44 age groups, respectively, for the urban sector in 1975–1984 and 1985–1994.*

number of woman-years was less than 40 in an age group, with the result that there are only five ASFR values for the urban-sector sample: 15–24 and 25–34 years age groups for the period 1975–1984, and 15–24, 25–34, and 35–44 years age groups in 1985–1994. There are five major observations. Firstly, the ASFR of the 15–24 years age group in the rural sector was less varied across the four periods. Secondly, the ASFR of the 25–34 years age group in the rural sector scarcely changed from 1955–1964 to 1965–1974 but sharply declined from 1965–1974 to 1975–1984. Thirdly, the ASFR of the 35–44 years age group in the rural sector was similar in the three earlier periods and declined sharply in 1985–1994. Fourthly, the ASFRs were higher in the urban sector than in the rural sector, except for the 15–24 years age group in 1985–1994. The final point is that the ASFR of the 35–44 years age group sharply declined in the rural sector from 1975–1984 to 1985–1994, but was maintained in the urban sector at a similar level to that in the rural sector in 1975–1984. Consequently, the total fertility rate (TFR) in the rural sector was 5.7 in 1955–1964 and 5.6 in 1965–1974, dropping markedly to 5.1 in 1975–1984 and 3.6 in 1985–1994. In contrast, the TFR of the urban sector was maintained at about 5.0 until 1985–1994, being similar to that of the rural sector in 1975–1984.

Many previous studies have shown that changes in economic conditions (Walsh 1987a) and in fertility and mortality patterns (Riley and Lehmann 1992) are much more prominent among rural to urban migrants than among rural sedentes. The comparison of fertility between sedentes and migrants of the same population in PNG (Umezaki and Ohtsuka 1998) and in other developing countries (e.g. Brockerhoff and Yang 1994; Goldstein and Goldstein 1981; Lee and Farber 1984) also demonstrated lowered fertility among migrants. The changing fertility patterns in the rural sector of Perelik population are unique because they show the opposite to other studies of populations in PNG.

Family planning practices

The main reason for fertility decline in the rural sector has been the success of family planning, which probably began around 1970. The impacts of sexually transmitted diseases (STD) on primary and secondary sterility rates is likely to have been low, compared to other populations in PNG (Jenkins 1993). According to a study of a population in New Ireland District (which has had similar historical and economic conditions to Manus Province) prior to independence, fertility was largely affected by the prevalence of gonorrhoea (Ring and Scragg 1973). The TFR values in New Ireland increased from 3.4 in 1952–1957, to 4.7 in 1957–1962 and 5.4 in 1962–67, due largely to the mass use of penicillin as a control measure against gonorrhoea since 1954. The TFR values for Perelik village were consistently high in 1955–1964 (5.7) and 1965–1974 (5.6), values which are almost identical to those of New Ireland District in 1962–1967. The consistently higher TFR in Perelik compared with New Ireland is due to the low prevalence, or absence, of gonorrhoea and other STDs in the former location. The authors' survey work indicates that there have been few patients suffering from these diseases in Perelik, at least in the decades since health services have been available.

Family planning practices have played the main role in the lowered fertility of women in Perelik village after 1975. As pointed out by Avue and Freeman (1991) and the Manus Provincial Government (1982), family planning has prevailed in this province. Tubal ligation has had a high uptake among grand multiparae (mothers with five or more children) until the mid-1980s. Health workers have promoted this with success among Manus women; convenience and the lack of charge or low cost for the procedure combined with an absence of side effects may account for its high acceptance rate. Health workers have promoted this with other family planning methods, such as oral pill, among Manus women (Avue and Freeman 1991). In contrast, the promotion of vasectomy and condom use has been less successful because males see birth control to be the responsibility of the female (Muirden 1982).

To increase the understanding of contraception uptake in Perelik across time, interviews with 24 Perelik women were conducted by the first author (YA) in 1997 about their family planning experiences. In general, uptake of modern contraception is likely to have begun around 1970, because one woman made it clear that she completed childbirth before 1970 without any information about family planning services being available. In the early stages of contraception uptake, the oral contraceptive pill was popular and available without charge, at Lorengau General Hospital. Between 1976 and 1984, three women who were aged between 30 and 38 years at that time underwent tubal ligation operations. Of the women interviewed, 16 of the total sample of 24 had experienced one or more of the following methods: oral contraceptive pill; injection (Depo-Provera or Depo-medroxyprogesterone acetate); sterilisation (tubal ligation); ovulation (rhythm); and traditional herbal medicine, although the last method was used by only four women. Altogether, five of the six women aged 17–30 years were users of modern contraceptives. Perelik women used a much wider variety of contraceptive methods than parturient mothers visiting Port Moresby General Hospital in 1990, a high proportion of whom used hormonal methods (pill and injection) (Klufio et al. 1995).

Why have urban Perelik women maintained high fertility while rural women have not? There are two plausible explanations. Firstly, urban Perelik male heads of households have been mostly public servants, teachers, and office workers and their earnings may have been adequate to rear many children. In fact, one of them stated during his temporary return to the village that the number of children he had was not an economic concern for him. Male attitudes toward the number of children they raise has been important in determining family size throughout PNG (Agyei 1984; Muirden 1982). Secondly, urban dwelling Perelik may have low interest in family planning because of a lack of information or lack of awareness of the need for it. Tulloch (1986), who studied family planning practices among women in Lae, found this to be the case in what is the second largest urban centre in PNG.

Carrying capacity on Manus, and migration

The final discussion focuses on an ecological factor, population pressure, and carrying capacity of the rural area in relation to population processes. According to the 1990 PNG census, mean population density was eight persons per square kilometre, varying markedly from less than five persons in many provinces to 70–120 persons per square kilometre in the provinces of the Highlands region (Riley and Lehman 1992; Ohtsuka 1994b). The population density in the provinces of the Islands region fell between these extremes, ranging from 6 to 17 persons per square kilometre, Manus Province having the highest population density. Baluan Island had much higher population density in 1990, 96 persons per square kilometer. Even this high number may be an underestimate of the real situation, the 'net' population density by the number of persons per arable land area being much higher because of the large proportion of non-cultivable land on this island.

In Perelik, slash-and-burn horticulture has been traditionally the most important subsistence activity. Village land has been traditionally owned, portion by portion, by 10 lineages, the members of each lineage usually sharing land for making gardens. In recent years, such sharing has declined, some lineage lands being exclusively used by one household of a lineage only. It is likely that population increase in recent times has made land insufficient, although it is likely that there has been some adaptation in the agricultural system in response to this pressure. Table 4.3 shows the number of gardens in use (excluding fallowed gardens) by the members of each Perelik lineage denoted by the letters A to J, according to whether they are using their own or borrowed land. The number of gardens in use, which were made on the land lent by Perelik lineages, are also shown. Lineages A and B had enough land for their own use and thus lent some portions to other lineages, whereas lineages D, E, F and G did not. Lineage C did not have enough land for its own use, but lent one garden to others. All the gardens made by lineages H and I were on land borrowed from other lineages, these being located not only in Perelik territory but also in other village territories. These land use

Table 4.3. *Number of gardens in use by all households of the 10 Perelik lineages, broken down by own land and borrowed land, and number of gardens in use by other households in the 10 lineages' land*

Lineage (sub-lineage)	Number of households	Total number of members	Number of gardens in use — Own land	Number of gardens in use — Borrowed land	Number of gardens in use by others
A	1	5	11	0	29
B	2	5	7	0	7
C	11	47	34	19	1
D	2	2	5	0	0
E	4	16	11	24	0
F	2	12	4	8	0
G	1	9	9	14	0
H	1	4	0	6	0
I	3	11	0	19	0
J*	0	0	–	–	–

* All members of this lineage were out of the Baluan Island in 1995.

patterns manifestly imply shortage of land for cultivation, with five of the nine lineages making the majority of their gardens on borrowed land. The population density of Perelik is likely to be at, or in excess of, the carrying capacity of the land, at existing levels of technology (Ohtsuka *et al.* 1995; cf. Harris 1978; Umezaki and Ohtsuka 2002).

There are a number of possible reasons for rural to urban migration. The attractiveness of a modernised lifestyle and possible opportunities for 'modern' occupations outside of subsistence cannot be denied. However, we argue that 'push' factors, in the form of high population density per unit of arable land has played as significant a role in out-migration as the 'pull' factors of urban living. The recent increase of migrants without specific purpose might reflect push factors; the increase of job opportunities in urban areas has been slow and the Perelik peoples' advantage in finding such jobs in earlier times is diminishing. In PNG, a prevailing way in which urban migrants without regular or formal-sector occupation can subsist is co-habitation in their *wantoks'* spontaneous settlements of same language-group migrants. Urban dwellers of some populations such as the Wosera Abelam of East Sepik Province (Curry and Koczberski 1998) and the Huli of the Southern Highlands Province (Umezaki and Ohtsuka 2003) are able to accommodate incoming *wantok* members in their own language-group settlements in Port Moresby and elsewhere.

However, Perelik patterns of urban residence are different. They are scattered across towns in individual dwellings, most of which are owned by their employers, and are thus unable to support new migrants in a way that custom would demand. In the absence of a fully-functional *wantok* system in the urban setting, it is unlikely that rural to urban migration will be able to absorb surplus rural Perelik population into the future, due to a limited capacity of the urban sector in PNG. Skeldon (1985) suggested that this may become true more generally in PNG.

At some stage in the future of the Perelik population, urban to rural migration may exceed that of rural to urban migration. As yet, there have been few return migrants, but most urban-dwelling Perelik people express intentions of returning to their homeland when they retire, a sentiment also expressed by many urban-dwelling people in PNG (Kane and Lucas 1986; Ross 1984). In the case of Perelik, many jobless urban dwellers may have to return to their rural place of origin simply because their dependence on relatives with salaried jobs may not be possible. There is a strong possibility that such a back-migration in the future will increase the population density of Perelik village, and will therefore place further pressure on the agricultural system. What the longer-term outcomes of this might be are not clear.

Summary and conclusion

The analysis of reconstructed demographic data for the Perelik population across the period 1955 to 1995 has highlighted three major findings. Firstly, the total Perelik population gradually increased until the mid-1970s, due largely to a decline in mortality. However, rapid population increase thereafter was caused by both high fertility and to some extent by marriage into Perelik. The large population increase in the period 1975–1984 was due mostly to further declines in mortality, especially child mortality, even though family planning practices have gradually prevailed to decrease fertility since 1970, particularly in the rural sector. Secondly, the increase of the whole Perelik population after 1985 was largely due to the high fertility in the urban sector and not the rural sector. This implies that rural to urban migration may have reduced population pressure in the rural sector, but may have contributed to greater increase in the combined rural and urban population than might have been the case if migration rates to urban centres were lower. Finally, the shortage of land in Perelik village might have contributed to greater uptake of family planning in the rural sector, as well as contributing to rural to urban migration, producing a large number of urban dwellers without jobs in recent years. Given the problems associated with urban residence for these people, it might be anticipated that back-migration may become a serious issue in the future. Rural development in Manus is therefore imperative to be able to sustain population densities which already seem to exceed the carrying capacity of the land, at current levels of technology.

Acknowledgements

The fieldwork, on which the present study is based, was financially supported by a Grant-in-Aid for Overseas Scientific Surveys from the Japanese Ministry of Education, Science and Culture. The authors are greatly indebted to Perelik people for their kind acceptance of, and cooperation in, the fieldwork.

References

Agyei, W.K.A. 1984. 'Fertility and Family Planning in Papua New Guinea', *Journal of Biosocial Science* 16: 323–334.

Ataka, Y. and R. Ohtsuka 2000. 'Resource Use of a Fishing Community on Baluan Island, Papua New Guinea: Comparison with a Neighboring Horticultural-fishing Community', *People and Culture in Oceania* 16: 123–134.

_____, _____, T. Inaoka, M. Kawabata, J. Ohashi, M. Matsushita, K. Tokunaga, S. Kano and M. Suzuki 2001. 'Variation in Malaria Endemicity in Relation to Microenvironmental Conditions in the Admiralty Islands, Papua New Guinea', *Asia-Pacific Journal of Public Health* 13: 85–90.

Avue, B. and P. Freeman 1991. 'Some Factors Affecting Acceptance of Family Planning in Manus', *Papua New Guinea Medical Journal* 34: 270–275.

Bogin, B. 1988. 'Rural-to-urban Migration'. In *Biological Aspects of Human Migration*, eds C.G.N. Mascie-Taylor and G.W. Lasker, 90–129. Cambridge: Cambridge University Press,.

Brockerhoff, M. and X. Yang 1994. 'Impact of Migration on Fertility in Sub-Saharan Africa', *Social Biology* 41: 19–43.

Carrier, J.G. and A.H. Carrier 1989. *Wage, Trade, and Exchange in Melanesia: A Manus Society in the Modern State*. Berkeley: University of California Press.

Curry, G. and G. Koczberski 1998. 'Migration and Circulation as a Way of Life for the Wosera Abelam of Papua New Guinea', *Asia Pacific Viewpoint* 39: 29–52.

Firth, S. 1983. *New Guinea under the Germans*. Carlton: Melbourne University Press.
Garnaut, R., M. Wright and R. Curtain 1977. *Employment, Incomes and Migration in Papua New Guinea Towns*, IASER Monograph 6. Boroko: Institute of Applied Social and Economic Research.
Goldstein, S. and A. Goldstein, 1981. 'The Impact of Migration on Fertility: an "Own Children" Analysis for Thailand', *Population Studies* 35: 265–284.
Harris, G.T. 1978. 'Responses to Population Pressure in the Papua New Guinea Highlands, 1957–1974', *Oceania* 48: 284–298.
Jenkins, C. 1993. 'Fertility and Infertility in Papua New Guinea', *American Journal of Human Biology* 5: 75–83.
Kane, P. and D. Lucas 1986. 'An Overview of South Pacific Population Problems', *Asia-Pacific Population Journal* 1: 3–16.
Klufio, C.A., A.B. Amoa and G. Kariwiga 1995. 'A Survey of Papua New Guinean Parturients at the Port Moresby General Hospital: Family Planning', *Journal of Biosocial Science* 27: 11–18.
Lee, B.S. and S.C. Farber 1984. 'Fertility Adaptation by Rural-urban Migrants in Developing Countries: the Case of Korea', *Population Studies* 38: 141–155.
Manus Provincial Government. 1982. *Reference Volume on Provincial Data*. Lorengau: Manus Provincial Government.
Mead, M. 1934. 'Kinship in the Admiralty Islands', *Anthropological Papers of the American Museum of Natural History* 34, pt. 2, New York: American Museum of Natural History.
Ministry of Health. 1994. *Handbook Health Statistics Papua New Guinea 1991*. Port Moresby: Ministry of Health.
Mitton, R.D. 1979. *The People of Manus*, Record No. 6, National Museum and Art Gallery, Port Moresby.
Muirden, N. 1982. 'Family Planning Programmes'. In *Population of Papua New Guinea*, ESCAP/SPC Country Monograph Series No. 7.2, 117–129. New York/Noumea: United Nations/South Pacific Commission.
National Statistical Office. 1994. *Report on the 1990 National Population and Housing Census in Manus Province*. Port Moresby: National Statistical Office.
Ohtsuka, R. 1986. 'Low Rate of Population Increase of the Gidra Papuans in the Past: a Genealogical-demographic Analysis', *American Journal of Physical Anthropology* 71: 13–23.
_____. 1994a. 'Genealogical-demographic Analysis of the Long-term Adaptation of a Human Population: Methodological Implications', *Anthropological Science* 102: 49–57.
_____. 1994b. 'Subsistence Ecology and Carrying Capacity in Two Papua New Guinea Populations', *Journal of Biosocial Science* 26: 395–407.
_____, T. Inaoka, M. Umezaki, N. Nakada and T. Abe 1995. 'Long-term Subsistence Adaptation to the Diversified Papua New Guinea

Environment: Human Ecological Assessments and Prospects', *Global Environmental Change* 5: 347–353.

Parkinson, A.D. 1974. 'Malaria in Papua New Guinea 1973', *Papua New Guinea Medical Journal* 17: 8–16.

Riley, I.D. and D. Lehmann 1992. 'The Demography of Papua New Guinea: Migration, Fertility, and Mortality Patterns'. In *Human Biology in Papua New Guinea*, eds R.D. Attenborough and M.P. Alpers, 67–92. Oxford: Oxford University Press.

Ring, A. and R. Scragg 1973. 'A Demographic and Social Study of Fertility in Rural New Guinea', *Journal of Biosocial Science* 5: 89–121.

Ross, A.C. 1984. *Migrants from Fifty Villages*, IASER Monograph 21, Boroko: Institute of Applied Social and Economic Research.

Schwartz, T. 1962. 'The Paliau Movement in the Admiralty Islands, 1946–1954', *Anthropological Papers of the American Museum of Natural History* 49, pt. 2, New York: American Museum of Natural History.

Skeldon, R. 1982a. 'Recent Urban Growth in Papua New Guinea'. In *Population of Papua New Guinea*, ESCAP/SPC Country Monograph Series No. 7.2, 101–116. New York/Noumea: United Nations/South Pacific Commission.

———. 1982b. 'Internal Migration'. In *Population of Papua New Guinea*, ESCAP/SPC Country Monograph Series No. 7.2, New York/Noumea: United Nations/South Pacific Commission, 77–100.

———. 1985. 'Population Pressure, Mobility, and Socio-economic Change in Mountainous Environments: Regions of Refuge in Comparative Perspective', *Mountain Research and Development* 5: 233–250.

Tulloch, A.L. 1986. 'Family Planning Attitudes and Action', *Papua New Guinea Medical Journal* 29: 153–156.

Umezaki, M. and R. Ohtsuka 1998. 'Impact of Rural-urban Migration on Fertility: a Population Ecology Analysis in the Kombio, Papua New Guinea', *Journal of Biosocial Science* 30: 411–422.

——— and ———. 2002. 'Changing Migration Patterns of the Huli in Papua New Guinea Highlands: a Genealogical-demographic Analysis', *Mountain Research and Development* 22: 256–262.

——— and ———. 2003. 'Adaptive Strategies of Highlands-origin Migrant Settlers in Port Moresby, Papua New Guinea', *Human Ecology* 31: 3–25.

Walsh, A.C. 1987a. 'On the Move: Migration, Urbanization and Development in Papua New Guinea', *Asia-Pacific Population Journal* 2: 21–40.

———. 1987b. *Migration and Urbanisation in Papua New Guinea: The 1980 Census*, Papua New Guinea Research Monograph No. 5, Port Moresby: National Statistical Office.

Ward, R.G. 1977. 'Internal Migration and Urbanisation in Papua New Guinea'. In *Change and Movement: Readings on Internal Migration in Papua New Guinea*, ed. R.J. May, 27–51. Canberra: Australian National University Press.

CHAPTER 5

Fertility and Social Reproduction in the Strickland-Bosavi Region

Monica Minnegal and Peter D. Dwyer

Individual persons die. Some of their morphological and physiological characteristics may be reproduced in their biological descendants. Some of their behaviours and ideas may be reproduced in other individuals. And so, too, the systems of relations – social and economic arrangements, kinship networks, mythical and symbolic understandings, built environments and so forth – within which they have participated and to which they have contributed may be reproduced long after they have departed. Here, then, the connotation of 'reproduction' is faithful to the etymology of the word: bringing into existence again, producing again, re-production.

But the word reproduction takes two primary meanings in academic literature. To most biologists its primary referent is the asexual or sexual production of first generation descendants of all organisms. This descriptive focus is combined with an interest in identifying the mechanisms of reproduction. And here, of course, for half a century, the emphasis has been upon genetic mechanisms and the ways these participate via biological reproduction in the transmission of particular characteristics – morphological, physiological, behavioural – across generations. Most social scientists would acknowledge this usage of the term reproduction though they might be sceptical – sometimes with good reason – of the certainty and generality attributed to deterministic causes and of the way in which notions of biological reproduction and evolution become conflated. But to social scientists the primary referent of reproduction is not that of the biologists. Rather, it is the meaning implicit in the previous paragraph. When social scientists write of reproduction their first concern is with the observation that form – morphological and physiological characteristics, behaviour, ideas and systems of relations – is often replicated in space and in time. Despite etymological considerations, biologists might use the word reproduction with reference to the production of descendants even if it happened that there was no correspondence of characteristics in the individuals of contiguous generations. By contrast, and with more attention to etymology, social scientists do use the word reproduction with reference to the replication of form even where biological mechanisms of inheritance and transmission play no part.

Given these two meanings attached to 'reproduction', it is fair to ask to what extent and in what ways they are connected in human systems. The perspective of most evolutionary ecologists – and of sociobiologists in particular – is of a necessary connection between the two, a connection in which social reproduction is contingent upon and, thus, ultimately subordinated to biological reproduction (e.g. Betzig 1997; Betzig *et al.* 1988; Gowaty 1997; Standen and Foley 1989). Our position is different. In this article we argue that connections between biological and social reproduction vary across societies and, in part, are correlated with forms of social structure. In the first instance we suggest that a move from more to less egalitarian systems of human relations is likely to be associated with a decoupling of social reproduction from biological reproduction. Our proposal is based on comparative data from two closely related and geographically contiguous Papua New Guinean societies. These data concern fertility, the subsistence contributions of children and subtle though important differences in marriage systems. We note, however, that questions of fertility were not a central interest at the times of our research and, further, because we worked with small communities our demographic data lack statistical power. In consequence supporting evidence for our proposal must be, in large part, anecdotal. For the purpose of this article our starting observation is simple. In one of those societies women had few surviving offspring. In the other, they often had many. In seeking to contextualize this difference we note Shih and Jenike's words: 'When significant differences in fertility and mortality exist between cultural groups that are either contiguous or coresident in the same territory, differences in cultural practices, history, and status may underlie differences in population processes' (2002: 21).

Kubo and Bedamuni: background

Kubo and Bedamuni people live in neighbouring areas of the foothills and plains of the Strickland-Bosavi region, along the southern edge of the central ranges of New Guinea (Figure 5.1). They are culturally, linguistically and technologically closely related (Dwyer *et al.* 1993; Kelly 1993; Knauft 1985; Minnegal and Dwyer 1998; Shaw 1996), relying for subsistence on bananas and tubers from gardens, sago flour from wild and managed stands of palms, and animals procured from forest and streams. Most Kubo territory lies between 80 and 200 m altitude, reaching

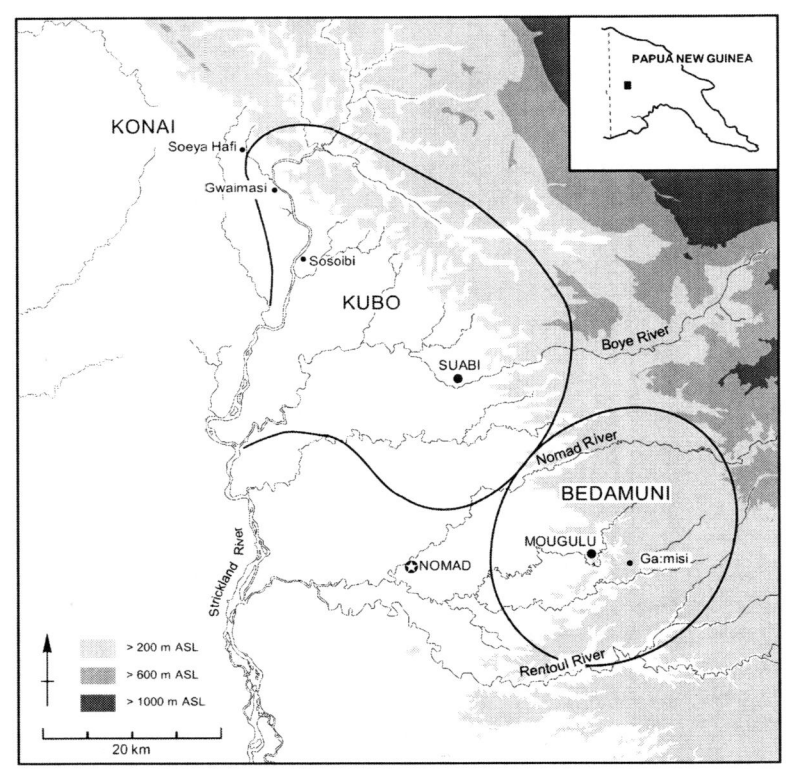

Figure 5.1. *Map of study area showing language groups and localities mentioned in the text. Suabi and Mougulu are mission stations within Kubo and Bedamuni territories respectively. Nomad is a government administrative centre*

to 400 m in foothills to the north and east, while most Bedamuni territory lies between 200 and 400 m altitude but reaches to 700 m and higher in the northeast.

Kubo number about 500 people, with an average population density of 0.4 per km², while Bedamuni number about 5,000 people, with population densities averaging 7 per km². Most Kubo live in widely dispersed communities of between 20 and 50 people, in contrast to Bedamuni communities of 100 or more people. And while Kubo communities tend to be flexible in composition, with people shifting residence frequently and freely, Bedamuni communities are generally stable with most men residing on their own clan land. Among Kubo, access to land is unrestricted, and association with land is established through its use (Minnegal and Dwyer 1999). Among Bedamuni, in contrast, access to land is inherited by men and provided to their spouses, and association with land can neither be established nor lost on the basis of performance. In both societies, marriage is predicated on exchange of siblings. While Kubo men frequently live with their wives' relatives, Bedamuni men rarely do so; Bedamuni women are expected to move to their husbands' community. As we have argued elsewhere, Bedamuni is a socially more complex society than Kubo, exhibiting 'increased differentiation within and between production units, greater integration within and between residential units, and heightened forms of evaluation within and between cultural systems' (Minnegal and Dwyer 1998: 375).

The present paper draws on research conducted in the Kubo community of Gwaimasi over 13 years between 1986 and 1999, and the Bedamuni community of Ga:misi over four years from 1995 to 1999. Because the former community was small we include data from the neighbouring villages of Sosoibi and Soeya Hafi; the latter village included both Kubo and Konai speakers.

Fertility

Figure 5.2 summarizes kinship connections, at January 1999, of residents of the Kubo and Bedamuni communities of Gwaimasi and Ga:misi, respectively. From Gwaimasi, but not from Ga:misi, we have detailed records of the births of children, both living and deceased, through at least 13 years. The number of living offspring per married woman was lower among Kubo (1.92 ± 0.38, n = 10 at Gwaimasi) than among Bedamuni (2.88 ± 0.41, n = 26 at Ga:misi). Samples are, however, small and not standardized

A - GWAIMASI

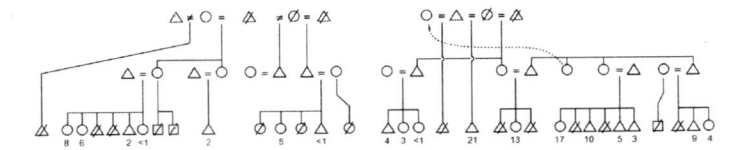

B - GA:MISI

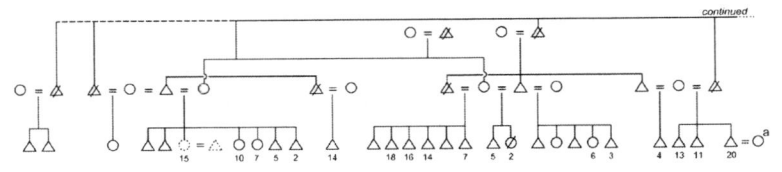

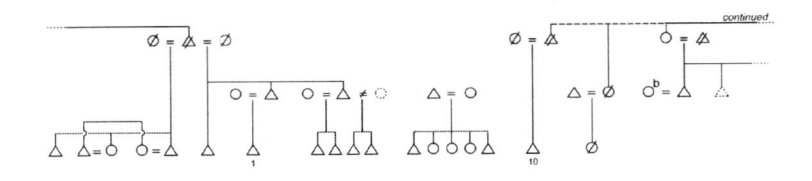

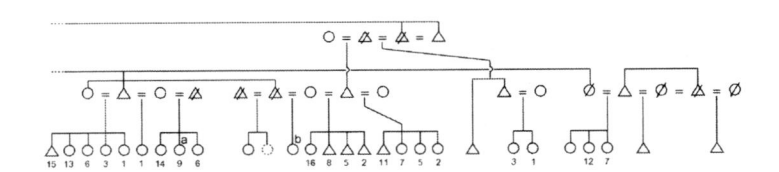

Figure 5.2. Kinship connections among residents of (A) the Kubo village of Gwaimasi and (B) the Bedamuni village of Ga:misi at January 1999. Available information has been simplified to highlight the composition of families; membership within particular clans, and other complex genealogical links, are not shown. Ages indicated for children are, for Kubo, based on our own records and, for Bedamuni, from records provided to parents by the hospital at Mougulu. On the Ga:misi diagram two females, indicated by 'a' and 'b', appear twice and dotted symbols indicate non-residents

with respect to age and reproductive lives of the women concerned. Table 5.1 explores available data in more detail.

For females over 20 years the number of living offspring was higher, in each age class, for Bedamuni than for Kubo, with the effect most marked for women older than 30 years. This pattern

Fertility and Social Reproduction 115

Table 5.1. *Fertility of Bedamuni and Kubo-Konai women*

Married females	Bedamuni	Kubo-Konai	Gwaimasi [Kubo]	
	Live offspring in Jan. 1999			Deceased offspring since 1984
Age class in 1999	Mean (range, n)	Mean (range, n)	Mean (range, n)	Mean (range, n)
8–20 years	0.25 (0–1, 4)	1.00 (0–2, 3)		
21–30 years	2.20 (1–4, 5)	1.78 (0–4, 9)	2.00 (0–4, 6)	1.50 (0–4, 6)
31–40 years	4.75 (1–7, 8)	2.43 (0–4, 7)	1.67 (0–4, 3)	1.67 (0–4, 3)
> 40 years	4.00 (3–7, 4)	2.57 (1–4, 7)	2.00 (2, 3)	0.33 (0–1, 3)

Only married females for whom the numbers of all living offspring are known are included. They are classed according to known or estimated age at January 1999. The Bedamuni sample comprises females from Ga:misi village. The Kubo sample comprises women from Gwaimasi village together with those from the villages of Sosoibi and Soeya Hafi. Data from Gwaimasi are shown separately both as number of live offspring in January 1999 and number of deceased offspring born since 1984. At Ga:misi, in January 1999, two girls, one 8 years old and the other 15 years old, participated in immediate exchange marriages. It was understood that the marriage of the younger girl would not be consummated until she had menstruated.

may be the result of either or both lower survival of Kubo children or a lower birth rate of Kubo women relative to Bedamuni. A high child mortality rate is apparent for Kubo. At Gwaimasi 14 of 29 children born after 1984 were dead by 1999 indicating a mortality rate to age 15 years above 48 percent.[1] No comparable data are available from Bedamuni, but some inferences can be drawn from data on average birth spacing in the two societies. In a sample of seven Kubo women, where data are adequate, the average interval between births resulting in living offspring at 1999 was 5.37 ± 1.75 years; for these women the average interval between all births was 2.67 ± 0.47 years. In a comparable sample of 10 Bedamuni women, the average interval between births of children still alive in 1999 was 3.26 ± 0.36 years. If these women, like those in the Kubo sample, gave birth on average every 2.67 years the expected number of births would have been 58.05. The observed number of living offspring from these women was 48. The two values would yield an estimated child mortality rate of 17.3 percent, a very low value relative to Kubo.

It seems probable, then, that both higher birth rate and higher child survival among Bedamuni contributed to the observed difference between the two societies.[2]

At least part of the difference in fertility profiles of the two populations may correlate with their different experiences of modernity.[3] The greater ease of access to medical care among Bedamuni, for example, may have contributed to lower child mortality rates. But availability of care does not explain why it is pursued, and nor does it explain why birth rates should be higher among Bedamuni than among Kubo. Other factors must have been implicated. In what follows, we first explore the contribution children make to potential biological and social reproduction in the two societies, before returning to a discussion of the relationship between these.

The work of children

Conventional accounts of cross-cultural variation in fertility rates among smaller-scale societies have argued that higher rates are facilitated by the combination of sedentism and an agricultural lifestyle (cf. Bentley *et al.* 1993; Sellen and Mace 1997). Kramer and Boone (2002) noted that recent review articles show this explanation to be incomplete in that, at the least, it fails to account for the fact that fertility rates of some horticulturalists may be lower than those of foragers. Drawing on data from a Mayan community (Mexico) they proposed that in as much as children contribute to subsistence, their work may subsidise the reproductive effort of parents rather than add to family wealth. According to their argument this effect is most pronounced where offspring produce more than they consume before the ages at which they marry and establish families of their own. In a comparative study of African foragers Blurton Jones *et al.* (1989, 1994) observed that Hadza children initiated productive work at a much earlier age than !Kung children (5 years versus 14 years) and concluded that, among the former people, foraging by children had reproductive benefits for mothers. They considered that the Hadza environment was less stressful than that occupied by !Kung and that this facilitated the difference in behaviour. In addition, however, Hadza foragers live at higher densities and range less widely than !Kung and, to this extent, may be utilizing land more intensively. Kramer and Boone's proposal is of interest here because, first, agricultural intensification is greater among Bedamuni than Kubo and, secondly, there are striking differences in the contributions made to subsistence by Kubo and Bedamuni children. These differences are expressed most strongly through the actual and expected contributions of girls.

Both Kubo and Bedamuni make banana gardens and tuber gardens but gardens of the latter type are more important and occupy proportionately larger areas and effort among Bedamuni than among Kubo (Minnegal and Dwyer 1998; 2001). From their mid-teens, boys in both societies contribute to both types of gardens and by their late teens are establishing and maintaining plots of the same size as those of older men. However, because the likelihood that teenaged boys and bachelors will have living parents is lower among Kubo than among Bedamuni, the distribution of garden produce is more likely to extend beyond the nuclear family among the former than the latter people. To this extent, unmarried Bedamuni males may directly subsidize the subsistence efforts of their parents to a greater extent than unmarried Kubo males. Further, where polygynous marriages have occurred – particularly those in which a man marries his deceased brother's wife – the age span of resident children within an extended family may be 20 years or more, providing opportunities for older unmarried sons to contribute substantially to the food requirements of their younger siblings. This effect too is much more pronounced among Bedamuni than Kubo (see Figure 5.2).

Many garden tasks are physically demanding – felling trees, clearing ground, fencing, carrying planting stock – and are beyond the capacity of young children. But once initial tasks have been completed, the work entailed in planting and maintaining tuber gardens is less arduous than that entailed at banana gardens. Bedamuni girls and a few Bedamuni boys maintain their own plots at tuber gardens. Nine plots attributed to children (8 girls, 1 boy) between 6 and 15 years old averaged 0.034 ha; this was 27 percent of the average size of 47 plots attributed to males and females older than 15 years of age. We have no comparative figures from Kubo because young children did not maintain discrete garden plots. Rather, they claimed particular plantings – a taro here, a clump of greens there – as their own.

The greater contributions of Bedamuni children to subsistence, and to subsidizing the work of mothers, is most evident with respect to processing sago from *Metroxylon* palms. In both societies most sago work is done by females (Dwyer and Minnegal 1995; Minnegal and Dwyer 1998). Among Kubo, girls rarely contribute until their early teens. When they are younger they are likely to accompany female kin to processing sites and may mimic the actions of adult workers by playing at miniature, flimsy troughs which have been made for them by the adults. We have only two records of a girl below the age of 13 years making sago. She was 10 years old at the time and produced flour at a rate of 2.50 kg/5 hours absence from the village on each of two days. This was 24 percent of the average yield achieved by married women older than 20 years ($n = 31$ records, 4 women). This girl did not work on any of eight other days when we weighed flour produced by her mother; indeed, her effort on those two days was probably encouraged by the fact that we had instituted a system of minor payment for the privilege of weighing sago flour at the end of each day's work. Among Kubo, even teenaged married females may produce less than experienced older women. Yields from 22 records by two young married females (19 and 20 years old, both with infants) were only 47 percent of those of older women ($t = 4.11$, $p < 0.001$). Qualitative observations suggest much variation in the yields of sago achieved by women of this age class. Young Kubo mothers may be often distracted from their work to provide care to infants. Older mothers are more likely to be accompanied by children who can attend to some of the needs of younger siblings.

By contrast with Kubo, Bedamuni girls as young as 8 years contribute to making sago, working at processing troughs which are built low to accommodate their height but which are otherwise of conventional design. Yields of sago flour produced by unmarried Bedamuni females less than 18 years old ($n = 9$ records from 8 females), were 87 percent of those of older married women ($n = 9$ records from 9 females; $t = 0.16$, not significant). Three of the unmarried females were 11, 12 and 13 years old respectively; their average returns from sago processing were, however, 95 percent of those of five unmarried females of 15 to 18 years in age. Thus, among Bedamuni, but not among Kubo, female children make a significant contribution to the production of sago flour. Their perceived importance as producers is reflected in the facts that far fewer girls than boys attend community schools located within Bedamuni territory and the expected immediate outcome of the marriage of a young, pre-pubescent girl is to increase the productivity of the household – that of the parents of the groom – which she joins.

Not only do Bedamuni children contribute more to subsistence – particularly household subsistence – but they are also less likely than Kubo children to inhibit the efficiency of their mothers' subsistence efforts. Bedamuni husbands and youths often remain at the village to care for young children while mothers are working at gardens. This was seldom observed among Kubo, where the division between work of men and women is less clear cut (Minnegal and Dwyer 2001).[4] In addition, the care of domestic pigs is more casual among Bedamuni than among Kubo. Indeed, among the latter people the demands upon a female carer's time through the first 18 months of a domestic pig's life are extreme, and concerns about risks to infants through contact with the urine or faeces of pigs means that the care of pigs is considered incompatible with late pregnancy and the first six months of lactation (Dwyer 1993; Dwyer and Minnegal in press).

It could thus be argued that Bedamuni children, more than Kubo children, enhance the potential for biological reproduction of their parents. But there seems to be no fundamental reason for the difference. Kubo children clearly could contribute more to subsistence than they do. That this is not asked or expected of them suggests that the value of children in this society lies elsewhere.

The value of children

To the extent that Bedamuni children contribute more to subsistence than do Kubo children, it might be thought that the former people would value children more highly than the latter people. The truth is more complex. Stated simply, where Kubo value all children of given gender equally – gender bias in favour of boys is sometimes apparent in both societies – Bedamuni value their own children more highly than the children of others. This is most evident with respect to the treatment of single parent children, orphans and otherwise disadvantaged children in the two societies.

In 1996–1997 at Ga:misi five children were either orphans or had one deceased parent. We exclude from the latter category children whose father had died but whose mother subsequently contracted a leviratic marriage with her dead husband's brother; these children were closely related to their current 'father'. The two orphans were 8 and 10 year-old males. The single parent children were an 8 year-old girl (father alive) and 12 and 17 year-old males (mothers alive but, in the latter case, not resident at the village). None of these children had older resident siblings. Though all had paternal or maternal uncles in the village, and associated most closely with the families of those relatives, they were not treated the same as their same-sex cousins. Relative to their age mates, the three younger boys were required to work hard, regularly accompanying their actual or surrogate mother when she went to gardens. They seldom joined the groups of young boys who spent many hours swimming, fishing or attempting to catch small game in the forest. Though the subsistence effort of the 17 year-old youth was not noticeably different from that of his age mates, in contrast to them there was no garden plot attributed to him alone. Two years later, in 1999, this now 19 year-old youth participated fully in the life of the community, with several banana and tuber plots of his own. In the latter period too, the now 12 year-old boy was better placed relative to his peers but life for the 14 year-old was precarious. It seems that incorporation of disadvantaged boys within a Bedamuni community is contingent upon proving their worth through a relatively arduous childhood subsidizing the reproductive potential of remaining or surrogate parents. An on-going difficulty for such boys, given the importance of marriage in forging and reinforcing alliances – and certainly at Ga:misi where in 1996–1997 there were more unmarried males than unmarried females (33 versus 23) – may be access to females as future marriage partners.

The single parent girl was in a more difficult position than were the boys. In part, this was because, at Ga:misi, her father was living in the natal community of his deceased wife and, from the time the latter had died, was judged to be and often treated as though he was an interloper. He was awkwardly placed in terms of both access to land for gardening and subsistence tasks that were usually the province of women. Neither he nor his daughter participated fully in sharing networks. The consequences for the girl were that she was expected to undertake tasks beyond her capacity and often to both support herself and contribute to her father's needs. Two extracts from field notes vividly capture her plight.

> We are nearly at the village ... and catch up with little Nebele, on her own, with a heavy load of bamboo shoots and a leaf package of unknown contents. There is no garden produce in her *bilum* [string bag]. She carries a bush knife and is bent over from the weight. She has been foraging for herself and her father today. He is at the village, he did not go out with her to help gather food. [21 January 1997. The girl's name has been changed.]

> Nebele's father carries four or five cooked bananas and at least one pitpit that he has been given. He sits on our verandah, the only person on one side of the two-part verandah but some children, including his daughter, on the other side. He eats by himself. Nebele goes across. She leans over the verandah less than a metre

from her father. She is begging for food but he ignores her. Finally he passes her one of the bananas. He does not look at her and no words are exchanged. [22 January 1997]

Food sharing among Bedamuni is hedged with more restraint than among Kubo and, outside formal occasions, is more likely to be confined to family members or close age mates. But even by Bedamuni standards Nebele's father's behaviour was extreme. He was himself placed in difficult circumstances and, in response, required that his daughter attend to her own needs. But those circumstances also meant that, unlike the boys discussed above, her contributions to the household did not subsidise potential reproduction. With the exception of a new born infant, Nebele was the only Ga:misi child to die in the extended drought that affected Bedamuni through much of 1997 (Minnegal and Dwyer 2000a).

There are no equivalent observations from Kubo despite the fact that our detailed records include eight unambiguous cases of orphans and six of single-parent children. None of these children had living older siblings, and yet all were fully accepted members of the communities at which they resided. Their needs for food and shelter were catered for without constraint. They were effectively 'adopted' by adults of their father's or mother's clans or even, if their widowed mother had remarried into a different clan, by adults of that clan. They were treated in the same way as same-sex classificatory siblings and were only minimally disadvantaged with respect to finding marriage partners.[5] Their actual or classificatory membership within a clan or set of 'brother' clans obviated any penalty that might have arisen from lack of direct parental support. Indeed, at the start of our research, our own recognition of their status as orphaned or single parent children came only when we compiled genealogical data. The single difference between these children and those with two living parents (who might not be both biological parents) is that the former moved, and were free to move, more than the latter from one subsistence unit or community to another. Perhaps more than most others, they gave expression to the ideal of mobility which informs Kubo ethos (Minnegal and Dwyer 2000b).

A sociobiological perspective on the preceding observations might emphasize the fact that Bedamuni appear to prioritize differentiation between children on the basis of biological relationship to a greater extent than do Kubo. An alternative, and perhaps more encompassing perspective, is that Kubo prioritize the incorporation of all children within a community-wide network of classificatory kin relationships to a greater extent than do

Bedamuni. On one count at least we can offer a rationale for this difference. Given the higher mortality rates of Kubo, with high rates of child mortality and with few children older than 15 years having both biological parents alive, the viability of the population as a whole may be contingent on maximizing opportunities for survival of all offspring. In the circumstance of a fragile demography, maximizing the survival chances of all children may help ensure that there will be future marriage partners for one's own children.

This argument shifts the emphasis of discussion from production of children to production of social relationships. A suggestion is emerging that, while Bedamuni value their own children for their potential contribution to the biological reproduction of kin, Kubo value all children for their potential contribution to the social reproduction of community. But Bedamuni children, too, reproduce social relationships. Something is still missing from our story. At this juncture we turn to the substance of our argument concerning the relationship between biological and social reproduction. Here we will draw upon differences in the marriage systems among Kubo and Bedamuni.

Decoupling biological and social reproduction

In an earlier article Dwyer (1996) argued that the shift from less to more intensified systems of production and from more to less egalitarian systems – in short, from less to more complex socioecological systems (see Minnegal and Dwyer 1998) – was correlated with changes in people's understandings of their own life-worlds. He proposed that, with increases in intensification and decreases in egalitarianism, people's understandings of 'culture' and 'nature' would be progressively decoupled. His particular focus was with relations between the visible (material) and invisible (spiritual) worlds as experienced by people from several Papua New Guinean societies. He proposed that as intensification proceeds people increasingly separate the two, ultimately peripheralizing (and reifying) either or both that which is understood to be sacred and that which is understood to be natural. He wrote:

> Intensification of production will simultaneously alter people's perceptions of both the visible and invisible world. At the outset a landscape in which use values are generalized, extensive and ungraded and in which invisible beings are all-pervading must be understood in totality as a landscape of human action and interac-

tion. Hence [for the people who live there], it is 'cultural'; there is no 'nature' and no contrast (Dwyer 1996: 178; the parenthetic insert has been added to the original).

Here we extend the reach of that earlier argument to explore relationships between social and biological reproduction. In particular, we suggest that where there is greater differentiation across the social landscape, with concomitant constraints on action and interaction, the processes that reproduce those social relationships will be increasingly separated from those entailed in producing descendants.

As discussed elsewhere (Dwyer and Minnegal 1999; Minnegal and Dwyer 1998; 1999), Bedamuni people differentiate more strongly than Kubo do between males and females, adults and children, and members of different clans. They place greater emphasis than Kubo on the integrity of local groups, and on the integration of those groups into a larger social landscape. And they, much more than Kubo, compare the behaviour and performance of individuals, categories and corporate groups and evaluate them accordingly. For Kubo, the differences between people are far more salient as a basis for social action than any similarities that might be recognized among them. For Bedamuni, relationships are established not just between individuals but between clans and social action is, accordingly, constrained at a higher, more embracing, level. In effect, and as elaborated below, the Bedamuni social landscape is strongly graded along several dimensions and, we will argue, the reproduction of those gradients is given priority over the reproduction of individuals in this society. The Kubo social landscape, in comparison, is relatively ungraded, and reproduction of individuals can be seen as simultaneously reproducing that landscape. The distinction is perhaps best illustrated by a comparison of marriage systems in the two societies, since marriage both facilitates biological reproduction and establishes social relationships.

As noted earlier, marriage among both Kubo and Bedamuni is predicated on exchange of siblings. But the two systems differ in important respects. The ideal Kubo marriage entails immediate exchange of sisters by two men who belong to different sets of 'brother' clans. A similar preference among Bedamuni is rendered more complex by the constraint that men should marry their father's sister's son's daughter (FZSD) or father's sister's son's son's daughter (FZSSD; Beek 1987: 40–41; Sørum nd). Rather than simply establishing a relationship between two men, Bedamuni marriages explicitly seek to reproduce the relationship between two lineages. One outcome of this constraint is that delayed exchanges are more likely to occur among Bedamuni than among Kubo. The difference is further emphasised by the fact that where, among Kubo, a female's status as exchange sister to a particular male is decided very early in her life, among Bedamuni, it is a female's status as future wife to a particular male that is decided very early in her life. Kubo adults, it seems, are concerned to establish the potential for children to negotiate future relationships as spouses, in-laws and parents on their own behalf. Bedamuni adults, in contrast, use children to negotiate relationships among themselves as wife-givers and wife-takers. In the Kubo case, females have greater powers of veto over proposed marriage arrangements, males are more constrained by the wishes of their sisters, and elopement is more common than occurs among Bedamuni.[6]

Among Kubo, to a much greater extent than among Bedamuni, men who have exchanged sisters are likely to co-reside, combine in subsistence ventures and assume special responsibilities toward each other's children. The last feature is of particular relevance, given the potential tension that exists between the facts of social kinship as expressed in patrilineality and the facts of biological kinship as expressed in the mother's brother/sister's son (MB-ZS) relationship.[7] Despite that potential, there is no area in which veiled aggression (ritualized transgression) characterizes the relationship between MB and ZS among Kubo. For them this relationship is warm, unthreatening and unambiguous.[8] It exists in concert with a system of land ownership under which asserted associations are expressed through patrifiliation but which, in practice, does not inhibit residence or access to resources. The same relationship, though acknowledged, holds far less significance among Bedamuni. Nor could it, for there is no early identification of a future exchange relationship between a particular brother-sister combination. Further, the fact that a Bedamuni woman has no right to use her natal clan's land means that brothers-in-law are less likely to co-reside and that opportunities for a man to provide material support to his sister's children are limited. The lack of ambivalence in MB-ZS relations among Kubo may be understood, perhaps, as an explicit expression of the fact that, for Kubo, understandings and expressions of biological and social reproduction are highly concordant. We would go further and hypothesize that ambiguities in the expression of this relationship are most likely to appear in contexts where people understand social reproduction to take precedence over biological reproduction.

Finally, among Bedamuni, marital discord is likely to elicit a rapid response from the wife's brothers, a response that borders on aggression towards the husband and his kin for their perceived failure to value the alliance the marriage represents. By contrast, among Kubo, the response of others to marital discord is characterized by embarrassment and studious avoidance of both parties; it is a matter for the spouses alone to resolve.

The marriage systems of Kubo and Bedamuni thus effectively reproduce, in the former case, an egocentric landscape of dyadic social relations and, in the latter case, a sociocentric landscape of differentiated but interrelated corporate groups in which men and women are differently positioned. But marriage also has implications for biological reproduction, not just in legitimising sexual relations but in its implications for access to the resources required to raise children.

Systems of marriage in which men engage in immediate exchange of sisters, in which the primary extra-familial bonds established are those between brothers-in-law, and in which both parties to each of those marriages acquire equal rights of access to their partner's clan land, have the outcome that for all four parties to a paired marriage the newly forged bonds of sociality simultaneously create equal opportunities for biological reproduction. No one is disadvantaged by the marriage with respect to either the wherewithal of subsistence or their standing within an essentially egalitarian community, a community in which the identity and performance of each individual in his or her own right is more highly valued than the identity and coherence of superordinate groups such as clans. Here, then, social reproduction and biological reproduction may be closely concordant. The former directly facilitates the latter.

By contrast, systems of marriage in which men engage in delayed exchange of women and/or choice of marriage partners is tightly constrained, in which the primary bonds established are those between wife-givers and wife-receivers rather than those between the participants to that marriage, and in which formal rights of access to land (in contrast to the use rights acquired by women) remain unmodified after marriage, have the outcome that the newly forged bonds of sociality do not correlate directly with opportunities for biological reproduction. A marriage affects the social identities of all members of the clans that are parties to the transaction, but the new spouses are particularly advantaged in terms of potential biological reproduction. In such systems, social reproduction and biological reproduction are decoupled, with the needs of the latter potentially subordinated to those of the former. Here, access to the wherewithal of subsistence and one's standing within a local community are dependent upon a socially sanctioned marriage that creates alliances which ramify far beyond the parties to that marriage. Biological reproduction is contingent upon conformity with prevailing structures of social reproduction but is not directly facilitated by social reproduction.

Differences in the marriage systems of Kubo and Bedamuni are in the direction of, but less extreme than, the contrasting systems depicted above. Those differences with respect to marriage and access to resources, together with over-arching differences in levels of both agricultural intensification and societal complexity (Minnegal and Dwyer 2001), are in accord with an expectation that biological and social reproduction will be more closely articulated among the former than among the latter people. What, then, are the implications for fertility in the particular case of Kubo and Bedamuni, and more generally, of decoupling these two modes of reproduction?

Implications for fertility

Where the social landscape comprises dyadic relationships generated through the actions and interactions of equivalent individuals, the production of a new individual itself replicates relationships. But where the social landscape comprises graded relationships between groups of individuals, production of a new individual in one of those groups cannot, in itself, replicate those gradations. We would expect, then, that biological reproduction, and the children that are the material expression of that reproduction, would be valued differently in such systems.

For Kubo, the birth of a child is simultaneously an act of social reproduction. The social identity of those who contributed to producing the child is changed by its birth, and the use of teknonyms (in which parents of child X are renamed as 'X-father' or 'X-mother') emphasises that change (Minnegal and Dwyer 1999). Individuals become parents, older siblings, mother's brothers of the child. And these new identities affect the way others interact with them. In turn, those interactions are directed at replicating family (or subsistence) units *per se* rather than being deeply concerned with patrilineal arrangements. This is consistent with an emphasis upon the social identity of individuals (rather than groups) as a basis for action, with an emphasis upon the future (rather than upon genealogical connections) in the constitution of personal relationships, and with an emphasis upon engagement (rather than upon prior convention) in establishing and affirming rights to land (Minnegal and Dwyer 1999). Autonomy of action at individual and family levels is valued by Kubo and has the outcome that differentiation by age, gender and status are minimally important in structuring social relations.

Among Kubo, therefore, social reproduction entails the production of children, and the provisioning of them with means to negotiate further relationships on their own behalf. Children are valued in their own right and not in relation to their structural position or expectations concerning their future potential to create alliances for others. And men are accorded social recognition through their life-long creative engagement with land. Thus, there are few structural constraints on social reproduction and none that is likely to exert a controlling influence over expressions of biological reproduction, certainly none that differentiate people, as individuals or groups, according to the numbers of children they have produced.[9] Rather, as we suggested above, biological and social reproduction are concordant; that which facilitates or constrains the former applies equally to the latter.

For Bedamuni, in contrast to Kubo, the birth of a child is not in itself sufficient to reproduce society for, while a child may ensure continuity of the lineage, reproduction of the wider social landscape is contingent on relationships negotiated with others through that child. The existence of the child does not affect the social identity of those who contributed to its production; that identity is defined, rather, in categorical terms of gender and clan affiliation, and is altered only through change in the relationships between those categories. And the interactions entailed here are directed at, and consistent with, replicating relationships between the patrifilial land-owning groups (clans) that secure resources for males, ensure genealogical continuity through the children born of marriage, and (though outside the brief of this article) sustain relations with the spirit world through the performance of seances and initiations (Beek 1987; Sørum 1980; nd). Acknowledgement of convention and authority are valued by Bedamuni and contribute importantly to structuring social relations according to age, gender and status.

Among Bedamuni, therefore, social reproduction entails production of children who may provide the means to negotiate relationships on behalf of the clan. Children are valued for their potential to be given as wives or to be the recipients of wives, and thus reproduce alliances between clans even as they ensure the reproduction of their natal or husband's clan. In such systems, the number of children an individual produces constrains the capacity of that individual, and thus the clan, to negotiate relationships with other groups. Children are thus valued, too, for their potential contribution to further biological reproduction of their parents.[10]

The value of Bedamuni girls is in part based on the subsistence work they may contribute to their mothers or mothers-in-law, and the value of Bedamuni boys is in part based on their membership within and expected future contribution to their patriline, while children of both genders have value in terms of their potential for reproducing relationships between clans. But in terms of these valuations, orphans, single-parent offspring and physically or mentally handicapped children jeopardize fulfilment of the desires of others whose understandings and performance are focused on social reproduction. The participation of such children as full members of groups is contingent upon contributing more than is expected of age mates, it is not granted to them automatically as of right. The likely outcomes are that many disadvantaged children will gain less access to the wherewithal of subsistence and, ultimately, have fewer opportunities to marry and have children.[11]

Finally, the differential opportunities of children within Bedamuni society direct attention to a likely differentiation among males who approach maturity. Here, again, it may be expected that a male's opportunities to father children in accordance with societal expectations will correlate with his capacity to participate within and contribute to the reproduction of the extensive network of social relations that constitute that society.[12] In the Bedamuni case this entails a secure position within a local land-owning group, unproblematic marital arrangements and the capacity to produce food sufficient to support his family and, from time to time, contribute to public feasts. These requirements are, of course, mutually reinforcing. At Ga:misi, in 1999, they were strikingly met by four males, each with two living wives, whose families included, respectively eight, nine, nine and 12 living children of whom seven, six, eight and six had been fathered by the man concerned (Figure 5.2(b)). The additional children were the offspring of either a deceased wife or a deceased brother. These men were exceptionally well placed in terms of multiple wives and the age span of children to satisfy Bedamuni desires with respect to social reproduction.

In socio-ecological systems in which biological and social reproduction are effectively decoupled, the ability of women to produce and provision children, the future reproductive potential of children and variation in the life-time reproductive output of men will be, to a large degree, shaped by prevailing processes and structures of social reproduction. This is congruent with, but adds to, Skinner's (1997: 68) argument, based on data from Eurasian agrarian societies, that 'family system norms ... exert substantial impact on the configuration and size of offspring sets'.

For Kubo, production of descendants defines the social identity of individuals, and the associated interactions – such as marriage – simultaneously facilitate reproduction of self and society. For Bedamuni, however, it is possession of ancestors that defines social identity; producing offspring, in this case, may well reproduce oneself, but will only reproduce society if marriages are arranged such that assemblages of ancestors from different lineages are replicated. The undifferentiated value accorded to children among Kubo, and the differential value accorded children – and adults on the basis of the number of children they produce – among Bedamuni, can be seen as correlates of the extent to which biological and social reproduction are coupled or decoupled in the two societies. And one consequence of the decoupling of these domains of reproduction in Bedamuni understandings, we assert, has been to promote higher average fertility and greater variation in its expression.

Concluding remarks

We have argued that the demographic processes and outcomes which characterize particular societies must be understood in terms of prevailing relationships between, and local understandings of, biological and social reproduction. In the Kubo and Bedamuni cases at least some differences of demography are influenced by the degree to which these two modes of reproduction have been decoupled. It is important to note here that we are not suggesting merely that social factors may act as primary constraints on, for example, fertility such that men of higher status may achieve greater reproductive success than men of lower status. We do not dispute that this will sometimes be the case (e.g. Hawkes *et al.* 2001). Rather, our argument asserts that there are circumstances in which people understand social reproduction as dissociated from biological reproduction, prioritize the former in their engagement with the world and, to the extent possible, pattern biological outcomes to satisfy social desires. In the particular case of Kubo and Bedamuni we identify those circumstances as associated with a decrease in egalitarianism and an increase in agricultural intensification and societal complexity.

We acknowledge that on a global scale the ecological and structural differences between Kubo and Bedamuni are relatively minor. But it was this very fact that, in the first instance, directed our attention to a major demographic difference between them and to ways in which the people understood and activated biological and social reproduction, respectively. Neither society should be taken as an extreme on a continuum of possible expressions of the entanglement of these two modes of reproduction. Indeed, we must allow also for variation within particular societies, variation which may be expressed, for example, in different understandings of, and responses to, the domains of social interaction which qualify respectively as 'inside' and 'outside' (cf. Godelier 1986: 173; Strathern 1991: 198). For example, among those Papua New Guinean societies where marriage arrangements are usually based in immediate exchange without bride wealth, exceptions are reported in which bride wealth is expected in cases of an unreciprocated marriage, a lengthy delay in reciprocation or a marriage contracted with 'outsiders' (e.g. Lemonnier 2001). Each of these circumstances reflects a heightened gradient in social interactions and a response that, in part, decouples biological from social reproduction. But perhaps decoupling is represented most completely where, for example, nation states advocate immigration in the interests of economic growth, scholars prioritize intellectual pursuits and publication over biological reproduction, and celibate religious orders are reproduced by training same-sex novices who have been 'captured' from the broader society within which those orders are situated. Arguments about demographic processes that are grounded in evolutionary ecology or sociobiology must always be partial unless they accommodate the fact that people may understand social and biological reproduction as being disconnected and may live in ways such that the latter is incidental to the former.

Notes

1. At the funeral of a Kubo infant the mother wailed and attempted to throw herself into the grave. She was held by other women who rebuked her strongly saying that she had living children, that they themselves and her mother in particular had lost many of their own children. Their message was that the death of children was the experience of all mothers and that while it was necessary to mourn losses it was more important to care for the dependent living.
2. Topographical considerations suggest that local exigencies of climate and parasite regimes are likely to be more stressful among Kubo than among Bedamuni. We acknowledge that these will contribute

to different mortality profiles and patterns of birth spacing in the two populations, but consider that causal factors are more complex and cannot be reduced to those of environment alone.
3. A missionary presence was established in Bedamuni territory, less than two hours' walk from Ga:misi, by the early 1970s and in Kubo territory, two days' walk from Gwaimasi, by the mid-1980s. In both areas this facilitated access to improved medical and health care which, presumably, contributed to the marked increase in population size observed among Bedamuni and the limited increase in population size among Kubo. Among Kubo, at least, cultural constraints on birth spacing were relaxed in response, the people said, to the teachings of pastors. In both cases, however, recent population increases have followed periods of extensive depopulation consequent upon epidemics of influenza and measles (cf. Kelly 1977; Shaw 1975). Beek (1987) reported that 39 percent of a sample of 174 Bedamuni were less than 15 years old in 1979. Our estimate from Ga:misi in 1997 is that 50 percent of the 95 Bedamuni residents were less than 15 years old. Comparable data from three Kubo-Konai communities give values of 31 percent of 78 people less than 15 years old in 1986 and 39 percent of 114 people less than 15 years old in 1999 (Minnegal and Dwyer 2000a). While at least part of the difference in recent fertility profiles of the two populations may correlate with their different experiences of modernity, those experiences do not, in themselves, explain what emerges from our data as striking differences in underlying patterns of fertility.
4. Bedamuni men spend less time than women in subsistence work, though this does not necessarily mean that they contribute less. The tasks they do undertake, especially the felling and fencing of gardens, require intensive effort over a few days, rather than the sustained work of maintaining gardens or processing sago. Kubo gardens, in contrast, require less preparation and less maintenance; men and women share much of the work at all stages of the gardening process, and their time commitments are thus similar. Also, Kubo men spend much more time hunting than Bedamuni men who have less access to forest and game.
5. What disadvantage there was resulted from the fact that exchange siblings – the 'sister' or 'brother' who will be exchanged for a spouse – are usually nominated soon after a birth. Thus, those who become identified as siblings at an older age may have to wait some time for an appropriate exchange sibling to become available. Where effective adoption crosses clan boundaries, however, the range of potential siblings may actually be enlarged; in more than one case the eventual exchange sibling was found in a step-father's clan.
6. Of course, demographic exigencies may disrupt the ideal of sibling exchange in both societies. Among Kubo such exigencies are often accommodated by fabricating a young female's status as exchange sister. Examples include: (1) a couple with three sons and one daughter who averted an expected infanticide by adopting an infant

girl born of an adulterous relationship; (2) the adoption by one man of his sister's young daughter as a future exchange sister to his own son; and (3) the adoption of a young girl, born of a (classificatory) incestuous relationship, as exchange sister to the son of the girl's mother's classificatory brother. The three adoptions were, in turn, rationalized as, first, grounded in fraternal kinship between the adulterous and adopting fathers, secondly, as return on an unreciprocated marriage exchange in the previous generation and, thirdly, given that the now widowed mother of the girl was establishing a marital relationship with another classificatory brother, as morally proper. In each of the cases the adopted girl was, at the least, classificatory kin to the father of her future exchange brother. An alternative strategy is for a man to exploit ambiguities in his own kinship to assert his status as brother to a young woman whose 'natural' exchange sibling has died. Comparable examples were not observed among Bedamuni. Rather, among these people a lack of actual sisters requires complex negotiations through which an exchange 'sister' is located among the daughters of the man's mother's classificatory sisters. Adoption of the girl by the parents of the man does not occur. This arrangement has the potential to enlarge access rights to gardening land of the girl's father and brothers in ways that would not be otherwise possible among Bedamuni. Access to land is not problematic for Kubo males who are free to garden on their own and their wife's land (Dwyer and Minnegal 1999).

7. Bloch and Sperber (2002) have recently drawn attention to earlier anthropological debates concerning the special relationship between mother's brother and sister's son that is evident, though variable in expression, in many patrilineal societies. With different emphases, contributors to those debates drew attention to the potential tension which exists between the facts of biological kinship as expressed in the MB-ZS relationship and the facts of social kinship as expressed in patrilineality. As Bloch and Sperber wrote: 'A cultural practice that acknowledges the rights of one's sister's children would normally go against the patrilineal norm and would be unlikely to stabilize ...' (2002: 733). Drawing upon an epidemiological approach to understanding representations they proposed that ritualized practices in which, for example, sister's son exerts rights over mother's brother's property may 'contribute to highlighting the normal character of patrilineal transmission of goods'.

8. On three counts the MB-ZS relationship among Kubo may be reinforced by conventional or ritual practices. First, in that from birth a female is declared to be the exchange sister to a particular brother. Secondly, in that when she marries and gives birth it is – or was in customary practice – her exchange brother and not her husband who undertakes a variety of practical tasks associated with preparation for the birth and support in the first few weeks after the birth. Indeed, if the woman is not resident on her clan land then she,

together with her brother and a retinue of female supporters may return to her natal land, leaving the husband behind. Or, if she does reside on her clan land, the husband is likely to depart when the birth is imminent and remain away for a week or more. Thirdly, at the time a young male is initiated his MB acts as his primary sponsor.

9. In an analysis of hunting success by Kubo males we were unable to detect a correlation with the numbers of offspring those men had fathered (Dwyer and Minnegal 1993). Nor has longer-term research contradicted this understanding.

10. In as much as additional children remain within the family unit so, as Kramer and Boone (2002) suggested, the effect on family size through underwriting parental effort may be extended. But this effect, we suggest, should not be taken as directly causal. It arises in the first instance because social reproduction both requires and provides opportunities for expressing particular demographic profiles.

11. The Baining provide an interesting case in which a married couple's adopted children are more highly valued than their own children because, in contrast to the latter, and as Fajans (1997) expressed it, they are 'made', they are the outcome of intentional productive actions by their adoptive parents. In this way they give concrete expression to the prioritization of social over biological reproduction.

12. Some early literature on highland societies of Papua New Guinea showed that Big Men were likely to have more wives and children than men of lesser status and that the sons of Big Men were more likely than the sons of other men to achieve Big Man status (e.g. Strathern 1971: 201, 209). While observations of these kinds will appeal to sociobiologists and, indeed, are congruent with sociobiological expectations they may also be understood as made possible in circumstances where inequalities emerge in association with the prioritization of social over biological reproduction.

References

Beek, A.G. van. 1987. The Way of All Flesh: Hunting and Ideology of the Bedamuni of the Great Papuan Plateau (Papua New Guinea). Ph.D. Thesis, University of Leiden, Leiden.

Bentley, G.R., G. Jasienska and T. Goldberg 1993. 'Is the Fertility of Agriculturalists Higher than that of Non-agriculturalists?' *Current Anthropology* 34: 778–785.

Betzig, L. ed. 1997. *Human Nature: a Critical Reader*. Oxford: Oxford University Press.

———, M. Borgerhoff Mulder and P. Turke (eds) 1988. *Human Reproductive Behaviour: A Darwinian Perspective*. Cambridge: Cambridge University Press.

Bloch, M. and D. Sperber 2002. 'Kinship and Evolved Psychological Dispositions: the Mother's Brother Controversy Reconsidered', *Current Anthropology* 4: 723–748.

Blurton Jones, N.G., K. Hawkes and J.F. O'Connell 1989. 'Modelling and Measuring Cost of Children in Two Foraging Societies'. In *Comparative Socioecology: The Behavioural Ecology of Humans and Other Mammals*, eds V. Standen and R.A. Foley, 367–390. London: Blackwell Scientific Publications.

_____, _____ and P. Draper 1994. 'Differences between Hadza and !Kung Children's Work: Original Affluence or Practical Reason?' In *Key Issues in Hunter-Gatherer Research*, eds E.S. Burch and L.J. Ellana, 189–215. Oxford: Berg.

Dwyer, P.D. 1993. 'The Production and Disposal of Pigs by Kubo People of Papua New Guinea', *Memoirs of the Queensland Museum* 33: 123–142.

_____. 1996. 'The Invention of Nature'. In *Redefining Nature: Ecology, Culture and Domestication*, eds R. Ellen and K. Fukui, 157–186. Oxford: Berg.

_____ and M. Minnegal 1993. 'Are Kubo Hunters "Showoffs"?' *Ethology and Sociobiology* 14: 53–70.

_____ and _____. 1995. 'Ownership, Individual Effort and the Organization of Labour among Kubo Sago Producers of Papua New Guinea', *Anthropological Science* 103: 91–104.

_____ and _____. 1999. 'The Transformation of Use Rights: A Comparison of Two Papua New Guinean Societies'. *Journal of Anthropological Research* 55: 361–383.

_____ and _____. in press. 'Person, Place or Pig: Animal Attachments and Human Transactions in New Guinea'. In *Animals in Person: Cultural Perspectives on Human-Animal Intimacies*, ed. J. Knight. Oxford: Berg.

_____, _____ and V. Woodyard. 1993. 'Konai, Febi and Kubo: the Northwest Corner of the Bosavi Language Family', *Canberra Anthropology* 16: 1–14.

Fajans, J. 1997. *They Make Themselves: Work and Play among the Baining of Papua New Guinea*. Chicago: Chicago University Press.

Godelier, M. 1986. *The Making of Great Men: Male Domination and Power Among the New Guinea Baruya*. Cambridge: Cambridge University Press.

Gowaty, P.A. ed. 1997. *Feminism and Evolutionary Biology*. New York: Chapman and Hall.

Hawkes, K., J.F. O'Connell and N.G. Blurton Jones 2001. 'Hunting and Nuclear Families: Some Lessons from the Hadza about Men's Work', *Current Anthropology* 42: 681–709.

Kelly, R.C. 1977. *Etoro Social Structure: A Study in Structural Contradiction*. Ann Arbor: University of Michigan Press.

_____.1993. *Constructing Inequality: The Fabrication of a Hierarchy of Virtue among the Etoro*. Ann Arbor: University of Michigan Press.

Knauft, B.M. 1985. 'Ritual Form and Permutation in New Guinea: Implications of Symbolic Process for Socio-political Evolution', *American Ethnologist* 12: 321–340.

Kramer, K.L. and J.L. Boone 2002. 'Why Intensive Agriculturalists have Higher Fertility: A Household Energy Budget Approach', *Current Anthropology* 43: 511–517.

Lemonnier, P. 2001. 'Women and Wealth in New Guinea'. In *People and Things: Social Mediation in Oceania*, eds M. Jeudy-Ballini and B. Juillerat, 103–121. Durham, Carolina: Academic Press.

Minnegal, M. and P.D. Dwyer 1998. 'Intensification and Social Complexity in the Interior Lowlands of Papua New Guinea: a Comparison of Bedamuni and Kubo', *Journal of Anthropological Archaeology* 17: 375–400.

_____ and _____. 1999. 'Re-reading Relationships: Changing Constructions of Identity among Kubo of Papua New Guinea', *Ethnology* 38: 59–80.

_____ and _____. 2000a. 'Responses to a Drought in the Interior Lowlands of Papua New Guinea: a Comparison of Bedamuni and Kubo-Konai', *Human Ecology* 28: 493–526.

_____ and _____. 2000b. 'A Sense of Community: Sedentary Nomads of the Interior Lowlands of Papua New Guinea', *People and Culture in Oceania* 16: 43–65.

_____ and _____. 2001. 'Intensification, Complexity and Evolution: Insights from the Strickland-Bosavi Region'. In *Agricultural Transformation and Intensification*, eds B. Allen, C. Ballard and E. Lowes, *Asia Pacific Viewpoint* 42: 269–285.

Sellen, D.W. and R. Mace 1997. 'Fertility and Mode of Subsistence: a Phylogenetic Analysis', *Current Anthropology* 38: 878–889.

Shaw, R.D. 1975. Samo Social Structure: a Socio-linguistic Approach to Understanding Interpersonal Relationships. Ph.D. Thesis, University of Papua New Guinea.

_____. 1996. *From Longhouse to Village: Samo Social Change*. New York: Harcourt Brace College Publishers.

Shih, C. and M.R. Jenike 2002. 'A Cultural-historical Perspective on the Depressed Fertility among the Matrilineal Moso in Southwest China', *Human Ecology* 30: 21–47.

Skinner, G.W. 1997. 'Family Systems and Demographic Processes'. In *Anthropological Demography: Toward a New Synthesis*, eds D.I. Kertzer and T. Fricke, 53–95. Chicago: University of Chicago Press.

Sørum, A. 1980. 'In Search of the Lost Soul: Bedamini Spirit Seances and Curing Rites', *Oceania* 50: 273–296.

_____. nd. *The Forked Branch: A Study of Meaning in Bedamini Ceremonial*.

Standen, V. and R.A. Foley, eds 1989. *Comparative Socioecology: The Behavioural Ecology of Humans and Other Mammals*. London: Blackwell Scientific Publications.

Strathern, A. 1971. *The Rope of Moka: Big-men and Ceremonial Exchange in Mount Hagen, New Guinea*. Cambridge: Cambridge University Press.

Strathern, M. 1991. 'One Man and Many Men'. In *Big Men and Great Men: Personifications of Power in Melanesia*, eds M. Godelier and M. Strathern, 197–214. Cambridge: Cambridge University Press.

CHAPTER 6

'EMPTINESS' AND COMPLEMENTARITY IN SUAU REPRODUCTIVE STRATEGIES

Melissa Demian

Adoption has long been recognised by anthropologists as a popular reproductive strategy in Oceania (Carroll 1970, Brady 1976), and Papua New Guinea (PNG) is no exception to this tendency. However, while adoption merits at least a mention in many ethnographies of Papua New Guinean societies, it has gone largely untheorised, as though its provisions were self-evident. Where discussions of adoption mechanisms do occur, they mainly appear as 'versions' of other exchange complexes, particularly in ethnographies of the Massim culture area (e.g. Young 1971, Weiner 1976, Chowning 1983), a part of New Guinea's Austronesian-speaking 'fringe' in which adoption is notably prevalent. Brief studies of adoption elsewhere in Papua New Guinea (Hogbin 1935–1936, Burridge 1959, Mandeville 1981) and in Melanesia (Keesing 1970) focused primarily on the classification of different adoption and fosterage forms, and on the identification of particular rights being transferred along with an adopted or fostered child. More recently adoption has been a component of the 'rediscovery' of the distinction between nature and culture (Fajans 1997), as if the Schneiderian critique of kinship theory and its ironically prolific aftermath had never occurred. Adoption appears, after all, to be the quintessential reproductive intervention, 'natural' relationships co-opted by 'cultural' imperatives. This is how adoption seems to have become intellectually crystallised in anthropology. For while kinship has by now been deconstructed and re-evaluated on nearly every other analytical front, adoptive kinship appears to be caught in a kind of time warp. It is not difficult to see why; even more so than marriage, adoption looks unassailably like a social strategy to manipulate biological relationships. Particularly in the West, where adoption is historically and stereotypically seen as the last refuge of the infertile and the orphaned,[1] it is seen as substituting explicitly for 'blood relations', a surrogate (as it were) for the children and parents who should have appeared 'naturally'.

One question which has usually gone unasked in anthropological considerations of adoption, Oceanic or otherwise, is what exactly is being *reproduced* by means of the adoption strategy. This question informed my own approach to adoption on the Suau Coast of Milne Bay Province, PNG. Much previous work on the topic has focused on adoption and fosterage as a 'house welfare' technique, either to consolidate or redirect the flow of property inheritance (particularly in land), or to negotiate relations between the adults involved in the adoption transaction. While elements of both of these perspectives may certainly be found in Western Suau[2] adoption, I chose instead to approach adoption as something like an aesthetic value (cf. Munn 1986). This decision was based partly on a desire to deprivilege 'kinship' as a rubric which seemed to close off rather than open up avenues of analysis, and partly on the strength of the social and cultural evidence in front of me. While adoption anywhere may be described fairly as a response to an undesirable reproductive state of affairs, of interest to the social anthropologist is what, precisely, makes this state of affairs undesirable. In Western Suau households, adoption can appear as much a solution to the reproduction of the wrong kinds of people as it is a solution to the failure to reproduce at all. Is adoption 'about' reproduction, then? If so, then it behoves us to ask what people think it is they are reproducing, whether by means of adoption or natalism.

During my fieldwork, I was constantly at pains to justify my focus on adoption as an ethnographic object. While it was as prevalent as I had anticipated – nearly every household had an adoptee in one of its generations – it was also so unremarkable that people often neglected to tell me who had adopted or been adopted even when I made this an explicit element of genealogical surveying. In the village of Leileiyafa, at least 12 percent of the population of 218 had been adopted or fostered;[3] because I was not always told who had been adopted, the actual prevalence may be higher. An adoption was not a high-profile event like marriage or mortuary transactions, which were categorically public activities, open to debate by the community, and of which people were generally able and willing to provide some exegesis. Adoption, by contrast, was more like childbirth in that it went uncelebrated and undiscussed.[4] It was resistant to singling out as 'an institution', because it always emerged from shifts in other, 'larger' relationships – those between husbands and wives, say, or between *bada* (mother's brothers) and *golowa* (sister's children). Adoption was, in my experience, most often caused to appear by other relationships operating upon, around, or because of it. My investigation of adoption became one of locating adoption in a social geography, that is, exploring the points at which it becomes a mapping-out of adjacent relationships.

One can therefore only talk 'about' adoption in the most literal sense of the word – that is, by describing the environment of relationships from which adoption emanates, a circumscription may be arrived at, a definition by means of negative space.[5] I invoke this English concept as a rough analogue to one in Suau: *'aha'aha*,[6] one possible gloss for which may be 'emptiness'. But *'aha'aha* does not always, or even usually, refer to a space which was once occupied but is now vacant. Rather, *'aha'aha* can refer to an incomplete image or a standard form which is missing an element needed to make it intelligible. In discussions of natalism with Suau informants, two phrases cropped up constantly: an 'emptiness of boys' (*tatau 'aha'ahana*) and an 'emptiness of girls' (*wawahin 'aha'ahana*). These terms indicated a sibling set composed of only one sex, a highly undesirable state of affairs. Grown daughters might berate their mothers for not providing them with any brothers; a couple with only boys may well view their situation with anxiety. Some people even ventured to tell me that having children of only one sex was not much better than having no children at all, a deeply pitiable condition for Suau. Doubtless these people were overstating the case, as there is a glaring disparity in material welfare between childless couples and those who have only boys or only girls, but the exaggeration served to make a point: no one would wish for this situation. While the desire for an equal number of male and female children has been documented in the literature on natalism in PNG (McDowell 1988: 31), it is worth investigating the contours of this desire surrounding the space a child should occupy in a particular social context.

Suau-speakers are matrilineal, like most other Massim societies, and normatively virilocal. The combination produces a residence pattern wherein many members of a matrilineage do not actually reside on land belonging to them, but they can and do maintain a 'long distance' interest in this land, often by making claims through their daughters. It would therefore seem obvious that people desire girls in order to maintain a stake in land belonging to their matrilineage. But is that the only 'function' served by girls? And what of the equal desire for boys; why is it just as calamitous only to have female as it is only to have male children?

Two families

Part of my approach to this question can best be illustrated by a pair of examples from Leileiyafa whose consideration in light of one another initially presented me with a conundrum. These examples concerned the two most prodigious adopting couples in the village, yet their reproductive, economic and social circumstances could not have been more different from each other. To begin with, a brief outline. Samalua and Elelia were middle-aged and therefore at the peak of their social influence, the senior couple residing at Ho'owatefana hamlet. Samalua's pig exchange network far outstripped that of any other man I knew, he was the only person I heard referred to by the honorific term *ba'isa*, and he belonged to a populous and powerful local lineage, Siwe'oya. He and Elelia had six hard-working natal sons who ran a small trade store and a speedboat service to the Commonwealth Development Corporation oil palm plantation at Sagarai. In addition to their four 'real' adoptees they had several casual, fosterage-type relationships with various children and adolescents who stayed in their hamlet and worked for them. These were, everyone agreed, 'strong' (*adidili*) people. By contrast, Joshua and Paini were in an extremely weak position. Elderly, frail, infertile, and living as guests on someone else's land at Lutulutudebadeba hamlet, they were often the recipients of other villagers' charity.[7] Nevertheless they had two adoptees and a fostered child.

One of the things that interested me was the fact that two couples in such starkly dissimilar circumstances were both multiple 'receivers' of other people's children. While it made good functionalist sense for an infertile couple like Joshua and Paini to adopt, it was less obvious why Samalua and Elelia had done so, since they were themselves so prolific. But each of these couples had arrived at adoption by different 'roads',[8] and so a view of their respective paths to adoptive parenthood had to be unfolded further. Adoption in these cases could not be taken solely as a measure of fertility, economic efficacy or land tenure strategising, but rather as an effect of the value of children and the way this value translates into action under particular reproductive conditions. In their turn, 'reproductive conditions' refer here to the anticipation of future relationships as much as, if not more than, they refer to relationships obtaining in the present. Suau adoption is an outcome of the desire to affect these future relationships as well as an outcome of the desire to affect relationships in the present. In concrete terms, an adoption may be contracted to 'cement' a new marriage (by enabling a young couple to 'practise' being parents) as well as to ensure that both mother and father will have helpers (*tausagu*) in their future productive work.

A breakdown of adoptions by Samalua and Elelia demonstrates the range of circumstances which brought their adopted children to them (Table 6.1). Unsurprisingly, only one of these conforms to a generally stated Suau ideal of adoption from a cross-sex sibling of one of the adopting couple. Christine's is also the only adoption which seems to have occurred for strictly 'ideological' reasons, that is, the correcting of a gender imbalance among their natal children, although Elelia told me initially that she had adopted Nancy for the same reasons. The story of her natal parents' jail sentence emerged much later. But this accounting of adoption is largely one of opportunism coupled with a well-established exchange network. A 'strong' household like Samalua and Elelia's can attract children others are unable to care for. Those who gave them children benefited by knowing the children would be well looked after, something they may have been unable or unwilling to do themselves. Samalua and Elelia benefited by engendering a debt[9] and by the obvious gain in the labour capacity of their family (especially for Elelia, who without adoption would have had no daughters to help her). Theirs was known as an industrious, prosperous family which relied on traditional forms of wielding influence; Samalua was conspicuously and consistently absent from involvement with village politics, but he was one of the first people anyone would go to if they needed a pig and his gardening prowess was widely noted. Unfortunately, the success of this family attracted as much jealousy as it did admiration and they were the target of sporadic robberies. The same high profile which brought them children occasionally brought them trouble.

The number and diligence of their natal children bore, however, no relationship to their propensity to adopt or who would inherit their land; although he had six sons, Samalua insisted that they would share the land with their adopted siblings. Samalua also took responsibility for the funerals of two of these additions

'Emptiness' and Complementarity 141

Table 6.1. *Samalua and Elelia's adoptees*

Name of adoptee	Relation to Samalua/Elelia	Details of adoption
Paulo	Samalua's MBSS	Samalua's bada (MB) had paid the bridewealth for a cross-cousin, but the cousin could not repay him in pigs and so he gave him a son, Paulo. When this man died Samalua took charge of Paulo, at the time about 12 years old.
Christine	Samalua's ZD	Christine was given to Samalua and Elelia by Samalua's sister as an infant because they had no female children.
Samoa	Elelia's FWBS	Elelia's father was a sorcerer who killed another sorcerer plaguing the Siwe'oya lineage (to which both Samoa and Samalua belong). To thank him they gave him Samoa, but he was too old to look after the boy and passed him on to Elelia.
Nancy	Elelia's FBDD	Nancy's parents went to jail for adultery when she was around ten, and as Elelia wanted another daughter she offered to adopt her.
Polo	? (may be none)	A somewhat unusual arrangement, apparently based on the close friendship between Polo and Samalua and Elelia's natal son, Kidi. Polo went to live in their hamlet for four years, and Samalua asserted that had Polo returned to his own hamlet, he would have demanded the compensation due an adopter.

to his family: Paulo, whom he buried on his land, and Polo, who had more or less 'adopted' Samalua because he was the closest friend and hunting partner of one of Samalua's sons. Polo lived and worked at Ho'owatefana and, after his untimely death, Samalua insisted on holding his funeral feast there, even though Polo had been buried on land belonging to his own lineage. Samalua was assiduous in carrying out the work expected of a father of these two young men, even though it was the work he might have anticipated they would one day perform for him.

When elaborated in a similar manner, the adoptions by Joshua and Paini appear less surprising than they did when the circumstances of the adopters alone were compared (Table 6.2). Again, only one of these is a 'conventional' adoption by Suau standards, and a common enough solution when one sibling has several children and another has none. Eroni maintained close ties to his natal family and said that he considered both Joshua and his (deceased) natal father as his 'father'. But Eroni's natal sister explained to me that his parentage would be identified by others, not by himself; that is, depending on where he travelled and which of his parents were better known in that place, people might refer to him as 'Yokotali's son', 'Joshua's adoptee', or, were he to go to the coast where his adoptive mother originated, 'Paini's adoptee'. It should be noted that none of the people adopted into this family were related to Paini, owing to her relatively distant village of origin. Following Paini's death in 1997, Eroni quit his job as an oil palm harvester and came with his wife to live with Joshua and look after him. In so doing he signalled the effectiveness of his adoption by doing what 'any' child would do, that is, reciprocating the parenting work that Joshua and Paini had invested in him.

The other two children were more casual arrangements. While Hamahewa was said to be 'not a real adoptee' due to the brevity

Table 6.2. *Joshua and Paini's adoptees*

Name of adoptee	Relation to Joshua/Paini	Details of adoption
Eroni	Joshua's ZS	Joshua's sister had failed to space her children properly and felt overburdened, so she gave him to Joshua and Paini to raise.
Lino	Joshua's ZDD (adopted)	An unusual adoption on top of an adoption. The couple on whose land Joshua and Paini were living had adopted her first, but when the wife of the family died the husband felt he could not look after Lino and gave her to his 'tenants'.
Hamahewa	Joshua's MBDS	More like fosterage than adoption in that Hamahewa came to stay with Joshua and Paini as an older child and eventually went back to his natal parents. No compensation was expected.

of his stay with Joshua and Paini, I could not help but note his temporary relocation to Joshua's hamlet after the death of Paini. Like Eroni, he returned to help Joshua both with the funeral and the resumption of daily life in its wake; unlike Eroni he eventually went back to his natal hamlet. Lino's first adoptive father (whose first wife had died) lived in the same hamlet as Joshua and Paini and she continued to divide her time between them. What is significant here is that there is an additional relationship between Joshua and Saulo, the hamlet landholder and original adopter of Lino. Joshua and Saulo's lineages stand in a 'funeral feast eater' (*tau'anban*) relationship to one another, which is why Joshua was invited to live on Saulo's land in the first place. But it is also possible that the 'road' of mortuary exchange creates a perpetual potential for other sorts of gifts to move between members of these lineages – including children.

Adoption transactions

The movement of children as gifts presents one of the most telling points of divergence between Suau adoption and its Western analogue. While only three people in Leileiyafa had been adopted in this way, the phenomenon was not singled out as being in any way remarkable. And in the past, I was told, children, under specific circumstances, were a prescribed type of gift – murder by sorcery could only be redressed by the giving of a baby girl by the kin of the sorcerer to the kin of his victim. The reason for this practice may have been, according to different people, that the girl's bride-wealth would serve as compensation for the death, or that she would one day bear children for the lineage that had lost a member to sorcery. As bride-wealth is a reciprocal obligation in Suau and the lineage affiliation of children is not altered by adoption, neither of these seems a particularly satisfactory explanation, but the specification of a girl in this transaction is not otherwise rationalised. The uncertainty surrounding this practice is understandable considering the decades that have elapsed since its currency. What matters here, I think, is simply the assertion that at one time, the gift of an infant was an acceptable means of directly transforming a negative relationship into a positive one. This is a property which is entirely beyond the scope of Euro-American adoption, which has instead evolved an ideal that the redistribution of children must be kept separate from economic (specifically, monetary) transactions.

The latter is a somewhat vexed notion. Although thousands of dollars can and do change hands in American adoption, for example, they are paid to the mediating agencies for services provided, and not the natal parents, to avoid any suspicion of the child having been 'bought'. This in turn generates a different set of problems, relocating value from the children to the parents. As Modell points out in her essay on fosterage and the plurality of family law in Hawai'i, '[I]f children are to be distinguished from a commodity, the conditions under which they are distributed must be couched in the language of well-being – a concept that instantly draws in value judgments' (1998: 160–161). Economics and kinship are ideologically distinct and incommensurate spheres for Americans, and so the adoption transaction is cast as being one of 'rights' rather than 'objects' (i.e. children). The appeal to a moral domain ('the best interests of the child') that Modell documents contains, among other things, the assumption that the 'interests' of children are served by a privileging of a certain type of parent to the moral and jural exclusion of others.

As the two cases from Leileiyafa demonstrate, this sort of picture is impossible to generate for Suau adoption. Adoption and its defining attributes 'disappear' if the actors directly involved in an adoption are the only objects of scrutiny. That is, drawing a distinction between 'child welfare' (O'Donovan 1993) on the one hand and 'house welfare' on the other (Watanabe 1963) is not helpful in a context where 'welfare' is conceived not as a set of material or psychological conditions, but as a set of *pre*conditions for the reproduction of appropriate relationships over time. As I have documented elsewhere (Demian 2000), the work done by children for parents is only partially material in nature. Its other component is metaphysical, in that children are held to reproduce the likeness or 'image' (*tautau*) of their parents, thereby constituting a living 'memorial' (*he'ihe'inoi*) of parents who have died. This is not, of course, to downplay the significant material benefit to having children, who will help their parents throughout their lives and prepare their bodies for burial upon death, as well as carrying their 'images' into the future.

It should be noted that all of these tasks are same-sex-based; a daughter is not her father's helpmeet, and a son cannot wash and dress his mother's corpse prior to burial. Following marriage, both same-sex and cross-sex relations come to bear, articulated about the *yohu* and *iha* relationships. *Yohu* are cross-sex siblings; elder *yohu* are accorded particular respect. *Iha* is the *yohu* of one's spouse, to whom one gives small gifts upon marriage, and to whom respect and assistance are owed in perpetuity (Figure 6.1). It is these relationships which, because they are dependent upon the presence of both sexes in a sibling set, contribute to the desire

'Emptiness' and Complementarity

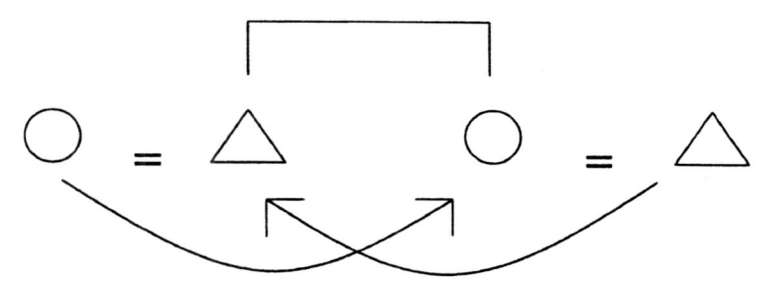

Figure 6.1. *Exchange between* iha

not only of parents for children of the same sex as themselves, but for their children to desire cross-sex siblings. A sibling set characterised by an 'emptiness of boys' or an 'emptiness of girls', then, is 'empty' because it does not permit the anticipation of adult relationships which depend on both same-sex and cross-sex axes of nurturance and support. Having children of only one sex may not literally be as calamitous as having no children at all, but the consequences are of a comparable gravity because the potential value of relationships with and among one's children will never be realised. When Samalua and Elelia began to adopt girls to rectify their 'emptiness of boys', they were pursuing an analogous strategy to that of Joshua and Paini who had no children of their own at all. To reproduce persons is not enough: one must reproduce the right kinds of persons in order to reproduce the right kinds of relationships.

Working for children and the work of children

To speak of 'kinds of persons', however, is to invite consideration of the ways that adopted children are distinguished from natal children. On the one hand, adoptees and their natal parents never 'lose' their relationship to one another, and so an adopted child has both rights and obligations stemming from his or her dual parentage. Within the adoptive family, there is ideally no distinction drawn between adopted children and natal children – save that they make the transaction of nurturing work between all children and their parents particularly explicit. A 'failed' adoption is one in which the child returns to his or her natal parents, whereas a 'successful' adoption is one in which the child lives with his or her adopted parents until marriage, and then continues to support them in their dotage.

The value of children is unromanticised yet critically important for Suau people, and not only for the obvious advantages of having children in a subsistence economy. A field assistant summed it up thus, in a conversation about the infertility of her uncle and his wife: 'With no children we just sit quietly all day, doing nothing, talking only to pigs and dogs.' This remark alerted me to the notion that it is children, not spouses, who are the ideal human companions.[10] Children are the trustees of their parents' memory and the embodiment of their positive relationships projected into the future. While not fully social themselves, children nonetheless render the adults around them social through their need for attention, instruction and nurturance (cf. Carsten 1991). What I mean by 'not fully social' is a specific imbalance in the exchanges between adults and children. Adults will in the same breath emphasise the overwhelming necessity of having children and the enormous amount of work having children entails; the work is 'worth it' because they can see the result in the visible growth of their children. It is the work which defines much of their own adulthood. And of course the children will one day work for them as well, both during and immediately following the death of their parents. But young children do no work – they are consumers only – and it is this debt which parents remind them of in their adolescence if they show reluctance to take on adult responsibilities.

Although infants are made much of, birth itself has a remarkably low profile. There may be something of a hubbub in cases where women go to the provincial capital to give birth in hospital and then return with their infants, but where women deliver in the village it may be days before more distant hamlets receive the news and longer before the baby's name is generally known. Stillbirths are buried perfunctorily and unmourned. They are not, however, 'inhuman'; rather, these babies may be said to have carried potential but unrealised relationships within them. When I returned to Leileiyafa on market day to learn the news of a stillborn baby girl, Glenda, the sister of the woman who had borne her, told me: 'Her skin was white like yours and her hair was straight like yours.' She then gave me one of the fish she was selling. I understood this gesture to mean: there might have been a relationship between your family and mine, but it did not come to fruition. The gift of the fish suggested a one-time substitute for what might have been a long-term exchange. I found myself choked with emotion even though I felt no particular affinity with Glenda's family. Her gesture had succeeded in causing me to feel the loss of a relationship which had not happened. The triangulation of Glenda, her sister's stillborn daughter, and myself was painful because of the absence of the daughter, the unrealised potentiality.

This is another, broader sense in which children are incomplete social beings. It could be argued that children are uniquely positioned to enable the sort of transaction I have just described: because immature, they hold within them the potential for all sorts of relationships which may or may not emerge. Adoption is one of these, and because it is only under certain circumstances that a particular 'kind' of child (male, female, orphan, infant, toddler) is singled out for adoption, *every* child has the potential to be an adoptee. The adoptive relationship is not so much created as 'found' or 'chosen' among the many other relationships possible to have with a child. Many relations of difference, and less often of sameness, must be deliberately created: marriage to a person of the opposite sex (which in turn creates *iha* of the same sex), funeral exchange with a person of a different lineage, adoption of another's child.

It is arguable that adoption may not be 'chosen' in the sense of an act of individual will; whenever I asked who had made the decision to adopt or give in adoption I was met with blank stares or, occasionally, a story involving two or more people with no indication of which of the people involved had decided that adoption was the solution to a particular need. Rather, adoption seems to precipitate out of the currents of social life almost of its own accord; it may be one of the 'built-in effects' which Strathern (1999: 65) has observed to concern the ways knowledge is to be acted on or how it might support or efface other knowledge. But whereas Strathern notes the way in which relationships appear to emerge fully-fledged from biogenetic information in Euro-American contexts, so that to reject the information is to reject the relationship, in the context of Suau adoption relationships are a product of recognition rather than revelation. Adoption is foregrounded by particular events, may 'disappear' into the background with the events themselves, and may not be invoked or built upon until the right set of circumstances emerges again. It may be helpful to look at the choice of adoption in another way. Rather than the elicitation of a particular relationship from a child, adoption may be seen as the un-choosing of other relationships. Strathern provides the following summary of this conceptualisation:

> Whether imagined as choosing from a pair of alternatives or choosing a single path out of many, the suppression of multiple possibilities in favour of one makes relationships visible through the

capacity of the self to activate them. At any point, action appears directed toward one specific relationship (the eliciting partner), which becomes the cause of the action itself ... An action implies an agent. Conversely, for the agent to 'appear' action must take a singular form (1988: 277).

If adoption is to be seen as one such form of action, then its choosing by several people at the same time, as my informants sometimes seemed to suggest, makes sense in that a 'collective' agreement between adoptive and natal parents should *ideally* have the appearance of a singular event. The *form* of adoption is singular, even though it involves multiple actors and multiple relationships. The event of a child's transference is easily located in time ('I was a small girl'; 'We got him straight after his mother gave birth') but not in intent. This is no accident. The intent in adoption, its agency, must appear to emerge from all the parties involved. Otherwise, they are not agents; the act of adoption does not 'belong' to them and therefore neither does the 'product' of the act – that is, the child-made-adoptee.

Strathern argues that the presentation of action as if singular in origin is necessary because of a distinction between 'agency' and 'cause' in Melanesian epistemology (1988: 273). An agent is defined as 'one who from his or her own vantage point acts with another's in mind. An agent appears as the turning point of relations, able to metamorphose one kind of person into another, a transformer' (1988: 272, emphasis removed). Cause, then, is the 'other' viewpoint which incites an agent to action. But there is an additional position as well: the agent may act 'because of' another, but also 'in reference to' a third person or group of people. Groups are collapsed into singularities not because their members are indistinguishable from one another, but because they occupy a common vantage point in relation to the agent. Thus Glenda gave me a fish *because* I had inquired about her sister's baby, but *in reference to* the baby and her putative physical similarity to me.

This kind of configuration may be seen to occur elsewhere. In marriage, the parents of the bride and groom exchange pigs with their children as referents, and the brothers and sisters of the married couple exchange money as new *iha*. Of course, bridewealth pigs also 'point to' the parents exchanging them. But the 'referents' in a transaction are not always members of the road of the transaction itself; that is, they may explicitly be a non-participant in the flow of wealth. This is true of funeral prestations in which the bereaved lineage exchange pigs with their *tau'anban* precisely because of the absence of a person who formerly participated in these relationships. The deceased person is the referent in these transactions and must deliberately be transacted 'around', that is, rendered invisible by means of their exclusion from the relationships put on display at a funeral feast. The transformations achieved in all these transactions are made manifest in the persons of their referents, who may not have participated in the transaction at all. Unrelated men and women are turned into husbands, wives and *iha*; a 'remembered' living person is turned into a 'forgotten' ancestor. The effects of these transactions are realised 'outside' of the transactions themselves, removed from the acts of giving and receiving. In order to see an effect, one must look anywhere *but* directly at the transaction.

Adoption transactions are no different from those cited above. However, I want to distinguish this claim from the idea of adoption as a 'transaction in kinship', because 'kinship' is not being moved from one parent to another – what is transacted is a child. Sometimes the child can do double duty as both object (the gift given) and referent (of the gift) in this transaction. The child-as-referent is transformed into an adoptee because the adoptive relationship is not created so much as witnessed or recognised by the parents and other adults involved. Natal parents and adoptive parents must act 'as one' *in reference to* the child being adopted and *because of* the witnessing village at large which will evaluate the appropriateness of the adoption process. Parents act with all their neighbours and relations in mind, and so must appear to be acting as a unit. When this unit shows signs of dissolving, as when an adoptee returns to his or her natal parents, the latter must compensate the adoptive parents to maintain the flow of positive relations between them. Compensation also 'reclaims' for the natal parents the work done for the adoptee by the adoptive parents; it relocates or displaces this work 'back' to the natal family. Because the compensation payment moves in the direction of the adopters, the work they have done for their child must accordingly move in the direction of the natal parents.

Agency and welfare

The foregoing discussion of agency would appear to suggest that while adopters do have the child 'in mind', adoptions are not necessarily made with the child's 'interests' in mind. This is not because of a disregard for the quality of a child's life. Far from it: rather, it is because the proper care of an adopted child is what demonstrates that the adoption has been done well. On the face of it, children do benefit. I found it significant that the two most highly-educated and well-employed people from Leileiyafa (neither of them resides there) had both been adopted as children. One of them, a high school teacher, told me in no uncertain terms that she accounted for her education by the fact that she had had two sets of parents supporting her. She also admitted, however, that she had accepted a job in a distant province because she felt incapable of meeting the demands now made of her by all her parents. She had had double the nurturance as a child, and now had double the reciprocal obligations as an adult. So the material benefits of adoption for children appear to cancel themselves out, but only insofar as they do for any child in Suau, since whatever parental work was done for the child must return to the parents when the child is grown.

The problem with the concept of 'child welfare', as O'Donovan (1993) has pointed out, is that it is defined and decided by adults. While this does not necessarily invalidate the intentions behind the concept, it calls into question how it is actually to be differentiated from 'house welfare', commonly understood to be the interests of the adopting family in the labour contribution of children or their perpetuation of its lineage. If child welfare is revealed to be based in actual practice on the competing claims of adults, what has happened to the children and their 'interests'? Put another way: can objects be agents? Do children, by virtue of their being moved between different households, lose their status as social actors? Do they have this status to begin with?

These questions are a rhetorical extension of O'Donovan's argument, and I raise them to illustrate the consistent problem with anthropological analyses of adoption. Where earlier writers were concerned with the comparative applicability of rights, I am concerned with the comparison of objects. While Suau parents determine the movements of children between households for a multitude of reasons, nearly all of them are to do with relations between adults, and the compensation paid in cases of failed adoptions is also between adults.[11] But it is by their very nature as objects of the adoption transaction that the unique potential of children as *persons* is realised. Most objects of transaction – shell wealth, pigs, money – may embody relations but cannot return the regard of those who gave them, and so cannot themselves become agents. Children, once grown, can; how they will 'regard' the parents who received them is dependent upon how *well* they were grown by these parents. An adoption is deemed successful if the adoptee acts the same way as natal children during the old age and upon the death of his or her adopted parent, because the grown adoptee was acted-upon the same way as natal children during their childhood. Parents who adopt and do not treat their adoptee properly run the risk of abandonment when the adoptee reaches adolescence.

I have made the claim that in Suau, children are incomplete social beings. One could argue that it is not even adulthood (as marked by marriage, residence in a new house, procreation) which completes them, but death. The consolidation of a person's memory, his or her memorialisation by descending generations, is the most 'complete' a person can ever be, in a milieu where a human lifetime represents a potentially infinite configuration of relationships, the selection, activation and maintenance of which delineate a methodology for living. Death renders these relationships finite; when someone ceases to be then their relationships must be 'finished' as well. Childhood marks the commencement of new relationships, although they may not be 'new' relationships in the sense of something unprecedented. Rather, they continue preexisting roads laid down by their ascendants. Children also do not manage their 'own' social networks at first, but learn how to do so through their interactions with adults, who are also nodes in these networks. In this sense they do not, perhaps, act as agents. The incompleteness of children necessitates an initial drawing-out of the relationships within them by (slightly more complete) adults. What this means is that the agency or non-agency of children can be seen from two viewpoints. If agency and cause are to be found in different persons, and agency is an essentially transformative capacity, then it can be located in different places depending on the 'status' of the adoption. An adoption entered into sees the parents as unitary agents, acting because of what I have called the 'witnesses', the body politic of the village, those who 'observe' the parents' relations to be in metaphysical or practical crisis. The adoptee is transformed by this process into the child of another set of parents, *who are, because of their position as agents, indistinguishable from the natal parents*. The idea can be crudely envisioned as in Figure 6.2.

This is, I believe, the 'key' (if there must be one) to understanding why Suau adoption generates new parent-child relationships with a facility that has little or nothing to do with the original 'status' of the adopting parents. The relationships appear as if 'not new', extant from the child's birth, because of the elision of the natal with the adoptive parents. But when an adoptee decides to return to her or his natal parents, the perspective shifts (Figure 6.3). The arrows are double-headed because the 'effect'

                                Parents
                                   a

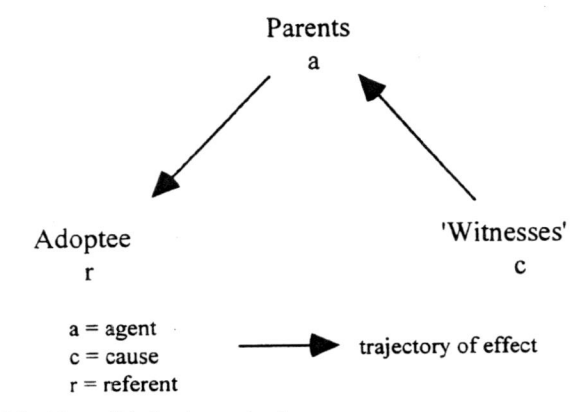

            Adoptee   'Witnesses'
               r   c

         a = agent
         c = cause   ⟶ trajectory of effect
         r = referent

Figure 6.2. *'Consolidating' an adoption*

may be perceived as originating in either set of parents – repulsion on the part of adoptive parents, attraction on the part of natal parents, or both – but what happens is that the adoptee's relocation divides the position of the parents. They can no longer be agents. The body of 'witnesses' should probably appear here as well, as their perceptions will have an impact on the actions of those involved; they are, perhaps, another 'cause'. But I have omitted them to show the effect of the returning adoptee on the parents who initiated the adoption. This is why adoptive parents must decide whether or not to demand compensation: the adoptee has chosen to differentiate between them, to dismantle the image of the unitary set of parents. This is the point at which the adopters, who have in essence been 'de-adopted' by their child, know their work on the child's behalf will not return to them.

Taken from the first vantage point in Figures 6.2 and 6.3 (parents as agents), do adoptive parents in Suau really see no difference between their adoptive and natal children? This is

                          Adoptive parents
                                 c

            Adoptee   ⟷   Natal parents
               a   r

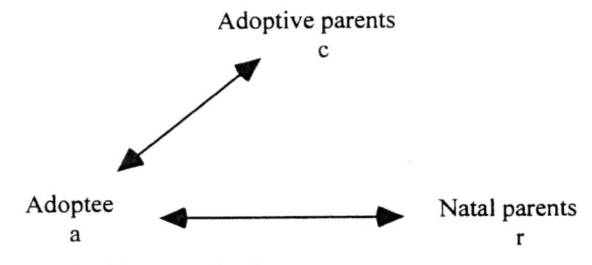

Figure 6.3. *'Dissolving' an adoption*

unlikely, as adoptive children still index their natal parents as well as their adopters, and so to regard them is to regard their natal parents 'behind' them. But the ideal of *acting* toward adopted children no differently from natal children creates the unified object of 'children' undifferentiated from each other. If the gardens built for them, the food fed to them, the clothes made for them, the money spent on them and the bride-wealth paid for them is the same, then they are constituted as no different from natal children because they are grown by the same work that natal children are. The 'view' of natal parents subsumed by adoptees' additional identity does not conflict with this identity, because natal parents have largely ceased to act on their children's behalf. Their relationship with the children is present, but dormant.

Suau adoption achieves an extremely specialised form of illusion. By this I do not mean to suggest one of Maine's 'legal fictions' (Maine 2000 [1861]: 13), but an optical effect. The transference of children between households generates an image of stillness out of movement, because a child is perceived to be 'rooted' in both places and because the transference itself creates a 'road', an unmoving connection, between them, along which other objects may move. Children who are not adopted cannot generate this effect because they 'end' in the same place that they 'begin'; they and their identity do not 'travel'. Adoptees, especially those whose natal lineage is different from that of their adopted lineage, extend the range of both by virtue of their movement. Although their options for marriage are narrowed, their options for land inheritance are broadened. Adoptees have more (or more of a certain kind of) potential relationships within them than their non-adopted peers. Adopted children simply make explicit what *all* children in Suau do implicitly. As gifts between adults, they carry out the objectification of relationships reproduced over time, the creative work required of each generation. Their movements back and forth between natal and adoptive households, and later their decisions about which family to align themselves with, are a visible manifestation of what everyone 'knows' about relationships in general: that they cannot be maintained but through the right work, and that they cannot be reproduced but through the products of work. Adoption is, then, not in itself a solution to the 'emptiness' of having reproduced incorrectly. It is merely the commencement of a possible solution, the completion of which will not be known until the end of the lives of those who initiated it in the first place.

The object(s) of reproduction

I would like to conclude with a few words on adoption as a kind of 'kinship'. Almost the entire analysis of adoption up to the 1970s concentrated on defining its 'jural' status cross-culturally, and then using the metaphor of commodity relations (with parental rights envisioned as a form of fungible property, e.g. in Goodenough 1970) to produce an analytical account of this status. The matter seemed to rest there, for very little attention was paid by anthropologists to adoption until the 1990s when Euro-American adoption came under scrutiny. A notable exception was, of course, that of Schneider (1984) who identified a tendency dating from the birth of the discipline for anthropologists to sideline adoption due to its non-biological status. He summarises the assumptions which early Melanesianists such as Malinowski and Rivers brought to adoption in their respective fields in terms of a genealogical method which could only recognise one kind of 'real' kinship, that is, kin connected by the 'bond of blood':

> [A]doption creates 'kinship' where none in fact exists, that is, no real blood relationship exists. Hence there ought to be a clear cultural distinction between true kinship and all other kinds of kinship.
> What is confusing is that adoption is confounded with the blood relationship by being called or treated as if it were the same kind of relationship. But in fact anthropologists have consistently treated adoption as something quite different from true kinship. For Maine, adoption was the first legal fiction, and as such allowed families to add new members by means other than birth. But there is no question that it is different from true kinship, blood relationship, the difference marked by the term *fictive* ... 'True' or 'real' kinship presumes that there *is* some biological relationship between persons so related (1984: 172, emphasis in original).

There is, in other words, a dual conundrum presented to the analysis of adoption, in Melanesia or anywhere else for that matter. The idea of adoption as a legal fiction has never quite been shaken from the anthropological imagination, but rather than being taken up by legal anthropology, adoption has somewhat inexplicably remained within the rubric of 'kinship'. But because it is not 'real' kinship (not only non-consanguineal, but non-affinal), anthropologists seem perennially at a loss for what to *do* with it. This conundrum is fundamentally one of being caught between comparative schemes. We have tried to have it both ways: adoption as a legal form, and also as a form of kinship, with the implicit understanding that these are nonetheless *mutually exclusive*. The problem is emergent from Euro-American adoption itself, whose peculiar embedding of a reproductive strategy in the judicial arena is a recent (early twentieth century) innovation and a vigorously contested territory (Modell 1994, 2002). At least partially as a consequence of this split identity, adoption chronically vanishes from the anthropological map of sociality. For anthropologists working in countries like PNG, which has a legal system that attempts to rationalise 'customary' practices like adoption with a body of statutory laws (Jessep and Luluaki 1994: ch. 6), adoptive relationships offer more than passing interest in the relationship between kinship and law. They require us to ask what adoption has to do either with kinship *or* with law, for to take either of these analytics for granted is to commit precisely the sort of ontological blunder of which Schneider accused twentieth-century kinship theory. It is, of course, arguable that both kinship and law are fundamentally concerned with the problem of reproduction, in that the former appears to reproduce the 'internal' capacities of persons, that is, their substance, while the latter appears to reproduce their 'external' capacities, that is, their relationships. Neither reproductive paradigm seems entirely suitable to adoption in societies of PNG, where what is being reproduced might just as easily be identified as temporality, aesthetic values, or indeed agency. I would therefore suggest that the task ahead is to allow the adoptive relationship to problematise our use of the domains of kinship and law, rather than to reproduce them uncritically. Far from being a footnote to considerations of 'real' kinship, exchange, or legality, the true potential of adoption lies in the way it might be used to turn some of social anthropology's most cherished rubrics inside-out, because of its ineluctable capacity to render explicit their conceits, rather than allowing them to operate unchallenged in the analytical background.

Notes

1. The latter belief is an interesting one, since there is no evidence that Western adoption ever drew solely or even mainly on a pool of orphaned children, but rather that it relied on children relinquished by poor and working-class women (see Zelizer 1985). The tenacity of the orphan myth suggests a powerful ideological necessity for natal parents to be 'absent' in order for adopters to be fully accorded the status of parenthood. One set of parents cannot exist in the presence (or knowledge?) of each other.
2. By 'Western Suau' I mean to indicate the half of the Suau Coast whose local appellation is Duiduileu, as well as its hinterland. Most

of the data in this article refers to field research conducted in the village of Leileiyafa in 1996–97. It should be noted that Leileiyafa, an inland village, is not considered by coastal dwellers to be a 'real' Suau village, since the dialect spoken there is more heavily influenced by inland than coastal variants of Suau. But neither is Leileiyafa a 'real' Buhutu or hinterland village, since it lies precisely at the point of convergence of the Daui (Western Suau) and Buhutu dialect groups.
3. Fostering, where children are given only temporarily to stay with another family, is much less common than adoption, which is understood to be 'for life'. But in neither case do children 'lose' their relationship with the natal family. A fostered child is likely to have weaker claims on inheritance than an adoptee.
4. Indeed, if one considers the post-partum seclusion formerly obligatory for a new mother, which would not have gone unnoticed by people in her hamlet, adoption could be said to be even less visible than childbirth because it was never accompanied by such observances.
5. I cannot claim this highly apt metaphor as my own contrivance, but had noted its appearance in Melanesianist writing (Strathern 1988: 11, Battaglia 1990: 140) and was perhaps 'primed' to look for it in my own fieldwork setting. While these writers use negative space to different epistemological ends – Strathern advocates a deliberate 'removal' of certain Western analytical conventions, and Battaglia finds negativities pre-existing in Sabarl evaluations of relationships – the fact that they are both influenced by Melanesian cosmologies is significant. Neither posits negativity as an absence, but more in the sense of an expectant space, waiting to be filled, which emphasises the relations and values around it.
6. The term would be rendered as *'afa'afa* in Leileiyafa dialect; I have given the standard Suau form.
7. This is meant very much in the Christian sense of the word. A member of the United Church Women's Fellowship in Leileiyafa once told me a story of the cleaning and repairing of Joshua and Paini's house when it had fallen into a pitiful state.
8. In the seemingly pan-Papua New Guinean sense of a relationship effected through exchange of some kind, such as marriage or compensation for a sorcery homicide. 'Road' can be glossed in Suau as *dobila* or *'eda*, and is used both to refer to physical tracks and to relationships, as in 'the road of marriage', *tawasola dobilana*.
9. While it may seem 'intuitive' to assume indebtedness to natal parents on the part of adopters, in fact the reverse is true according to Suau sensibilities. Adopters, in shouldering the 'weight/responsibility' (*polohe*) of raising a child, may claim compensation from natal parents if the adoptee returns to them.
10. A corollary to this is the fact that while unmarried women with children are said to 'lack respect' for their parents, they are not the objects of pity and bewilderment that married couples without

children are. The children of these women are wholeheartedly accepted into and cared for by their matrilineage.
11. Even cases such as one where a girl was adopted out because she was being neglected by her mother, had its root in the marital strife between her natal parents. The girl, a firstborn, was repeatedly abandoned by her mother whenever her parents fought, so her father exercised his firstborn prerogative and gave her to his sister. This does not mean the adoption was agreed upon 'unilaterally', as the sister and her husband had no children of their own at the time and were keen to adopt. What happened was that the father who gave his daughter away had to act 'as if' both he and his wife had agreed to the adoption, as is the stated ideal.

References

Battaglia, D. 1990. *On the Bones of the Serpent: Person, Memory, and Mortality in Sabarl Island Society*. Chicago: University of Chicago Press.
Brady, I., ed., 1976. *Transactions in Kinship: Adoption and Fosterage in Oceania*. ASAO Monograph 4. Honolulu: University of Hawaii Press.
Burridge, K.O.L. 1959. 'Adoption in Tangu', *Oceania* 29: 185–199.
Carroll, V., ed., 1970. *Adoption in Eastern Oceania*. ASAO Monograph 1. Honolulu: University of Hawaii Press.
Carsten, J. 1991. 'Children in Between: Fostering and the Process of Kinship on Pulau Langkawi, Malaysia'. *Man (N.S.)* 26: 425–443.
Chowning, A. 1983. 'Wealth and Exchange among the Molima of Fergusson Island'. In Leach, J.W. and Leach, E., eds., *The Kula: New Perspectives on Massim Exchange*. Cambridge: Cambridge University Press.
Demian, M. 2000. 'Longing for Completion: Toward an Aesthetics of Work in Suau', *Oceania* 71: 94–109.
Fajans, J. 1997. *They Make Themselves: Work and Play among the Baining of Papua New Guinea*. Chicago: University of Chicago Press.
Goodenough, W.H. 1970. 'Transactions in Parenthood'. In Carroll, V., ed., *Adoption in Eastern Oceania*. ASAO Monograph 1. Honolulu: University of Hawaii Press.
Hogbin, H.I. 1935–36. 'Adoption in Wogeo', *Journal of the Polynesian Society* 44: 208–215; 45: 17–38.
Jessep, O. and J. Luluaki 1994. *Principles of Family Law in Papua New Guinea*. 2nd edn. Waigani: University of Papua New Guinea Press.
Keesing, R.M. 1970. 'Kwaio Fosterage', *American Anthropologist* 72: 991–1019.
Maine, H. 2000 [1861]. *Ancient Law*. Washington, DC: Beard Books.
Mandeville, E. 1981. 'Kamano Adoption', *Ethnology* 20: 229–244.
McDowell, N. 1988. 'Introduction'. In McDowell, N., ed., *Reproductive Decision Making and the Value of Children in Rural Papua New Guinea*. IASER Monograph No. 27. Port Moresby: PNG Institute of Applied Social and Economic Research.

Modell, J.S. 1994. *Kinship with Strangers: Adoption and Interpretations of Kinship in American Culture*. Berkeley: University of California Press.

———. 1998. 'Rights to the Children: Foster Care and Social Reproduction in Hawai'i'. In S. Franklin and H. Ragoné, eds, *Reproducing Reproduction: Kinship, Power, and Technological Innovation*. Philadelphia: University of Pennsylvania Press.

———. 2002. *A Sealed and Secret Kinship: the Culture of Policy and Practices in American Adoption*. New York: Berghahn Books.

Munn, N.D. 1986. *The Fame of Gawa: a Symbolic Study of Value Transformation in a Massim (Papua New Guinea) Society*. Durham, NC: Duke University Press.

O'Donovan, K. 1993. *Family Law Matters*. London: Pluto Press.

Schneider, D.M. 1984. *A Critique of the Study of Kinship*. Ann Arbor: University of Michigan Press.

Strathern, M. 1988. *The Gender of the Gift*. Berkeley: University of California Press.

———. 1999. *Property, Substance and Effect: Anthropological Essays on Persons and Things*. London: Academic Press.

Watanabe, Y. 1963. 'The Family and the Law: The Individualistic Premise and Modern Japanese Family Law'. In A.T. von Mehren, ed., *Law in Japan: The Legal Order in a Changing Society*. Cambridge, MA: Harvard University Press.

Weiner, A.B. 1976. *Women of Value, Men of Renown: New Perspectives in Trobriand Exchange*. Austin: University of Texas Press.

Young, M. 1971. *Fighting with Food: Leadership, Values and Social Control in a Massim Society*. Cambridge: Cambridge University Press.

Zelizer, V.A. 1985. *Pricing the Priceless Child: the Changing Social Value of Children*. New York: Basic Books.

CHAPTER 7

Cognitive Aspects of Fertility and Reproduction in Lak, New Ireland

Sean Kingston

Some time ago, Mosko (1983) coined the term 'de-conception' to describe the Mekeo explanation of the way their mortuary rituals disarticulate the social relations between kin groups by unmixing the bloods that were combined to give bodily form to the person on their conception. The term has been taken up by other ethnographers and, particularly in Austronesian areas of Melanesia, usefully highlights the life cycle of the person and body as the scene for the making and unmaking of social relationships. Though we tend to forget, within 'Euro-American' culture at least, the meaning of conception, and hence also the implications of any neologism playing upon its reversal, extends further than the instigation of new bodies and lives by the physical and social intertwining of life's current incumbents. Though we seldom make the link explicitly, conception refers to the activity taking place in the head as much as in the bed.

There is a process by which persons come to cognitive, and social, presence, which is obviously connected to, though distinct from, their physical coming-into-being. However, it is the cognitive correspondents of de-conception that have thus far been highlighted as prominent features of discourse in Island Melanesia, and have been explored to great effect by a number of anthropologists (e.g. Battaglia 1990, 1992; Kingston 1998, 2003;

Küchler 1988, 2002; Maschio 1994) who take seriously informants' stress on the cognitive work to be done on the dead, which is often an overt aim of funerary ritual. This work is often expressed in terms of memory, of a forgetting or unthinking of the dead in favour of the living. This cognitive disarticulation is frequently associated both with the disarticulation of the social relations which had their nexus in the deceased, and with the physical disarticulation of the body and focal artefacts such as effigies.

However, surprisingly little attention has been paid to the origination of the person as a social and therefore cognitive event in Melanesia (though see Kingston 2003). The transformation of social relations that results in fertility requires comparable cognitive work to the transformation associated with mortality, and given the conception and deconception cycle of birth and death suggested by Mosko (1983, 1989) to be common throughout Austronesia (or in Strathern's (1988) terms of composition and decomposition, throughout Melanesia), one would perhaps expect the discourse of fertility to be as explicit as that of mortality, and expressed in similar terms. Rebirth, if not 'reconception', is after all also a primary metaphor in the male secret initiatory societies common in Melanesia (see Allen 1981), but perhaps the lack of investigation of corresponding (to mortuary usages) cognitive aspects of fertility is simply a consequence of the lesser attention paid to 'female' rituals in Melanesia altogether (for discussion of which, and degree of correction, see Lutkehaus and Roscoe 1995). It may be that idioms of substance have been overly concentrated upon (see argument of Thomas 1999), perhaps because of our own 'physicalist' understandings of bodies and their processes. However, while transmission and sharing of 'substance' is certainly important in Melanesia, such substances are neither the only important symbolic components for reproduction, nor are they distinguished from social or cognitive realms as they would be in the West.

Certainly in my field-site in the Lak[1] area of south-east New Ireland, a cognitive discourse relates to much more than mortuary practices. In fact, sociality itself is spoken of in terms of cognition. They give an explicitly performative dimension to the constitution of persons and their relations: for them social relations must be brought to cognitive prominence in order to take effect and there is an emphasis on the power of the person, artefact or performance to create, direct, modify and extinguish thoughts. For instance, a gift causes the recipient to think of one, and giving the gift is spoken of as 'thinking of' or 'remembering' (*namnai*) the recipient. Failing to heed the presumed relationship is, as one might expect, spoken of as 'not thinking' of someone or forgetting them. An 'associational', 'thoughtful' mode of sociality and identity formation is based on histories of interaction with persons, practices, objects and places. Social life is negotiated in an emotionally loaded and relational topography. The thoughts that items in this topography will inspire – turning the mind's eye towards this or that person or activity, exciting envy, shame, lust or duty – are prominent motivations attributed to actions in all spheres of life. One man will sit in view of his brother's house so that he will be thought of and fed; another will hide a laplap given by a friend so that his thoughts do not turn painfully to his absence; passing a clump of bush, another will reaffirm your relationship by reminding you how you cut some posts together there.

The object of many of these recognized modes of 'cognitive' action, if not ultimately all of them, is to affect the social constitution of persons. The transitions of the life cycle, in particular, are associated with rituals overtly directed at manipulating thoughts and attention in the cause of constituting and deconstituting persons.

A cognitive biology

Fertility and reproduction gain much of their character as topics for academic consideration from their centrality within the field of biology. A standard view is that biologies – including indigenous understandings of processes we deem biological – consist of various systems of explaining the material substrate of the person. The primary definition of biology in the *Oxford English Dictionary* is, after all, 'The science of *physical* life' (emphasis mine). Biological explanations on this kind of basis, in terms of substances, have certainly been prominent in Melanesia (see Knauft 1989 for survey), where the control of semen and blood, and their symbolic equivalents, often seems key to both personal and societal reproduction – such explanations no doubt reflect indigenous preoccupations as much as those of ethnographers.

Within Euro-American culture this equation of the biological with the physical also brings with it the ideological weight of objectivity. So, to suggest that fertility and reproduction might be understood cognitively seems counter to the biological project of physical explanation, and akin to supposing that the mental might explain the material or the subjective account for the objective – the reverse of the causal linkage normally admissible.[2] As the 'bodily' transformations given cognitive emphasis within Lak are physically and 'substantially' construed elsewhere, both in the West and Melanesia, how are they to be related to biology?

The answer lies perhaps within a subsidiary, though long-standing (e.g. Thompson 1961 [1917]), strand within Western biology that has gained new relevance in the informationally biased epoch of genetics. This understands biology and life processes through the lens of organization, of form, rather than of substances or materiality *per se*. The root of this conception is the isolation of reproduction as the formal basis of life. The prioritising of form over substance has gained such acceptance that previously unthinkable disciplinary constructions such as Artificial Life – the study and creation of self-replicating computer programmes, 'life in silico', as an exercise in 'synthetic biology' – are now pursued by academic institutions of high status and resources (see Helmreich 1998). This conception has gained particular prominence in modern science's focus on the micro-levels of biology, at which human development consists of differentiation of cells. Differentiation is biologically defined as a progressive developmental change to a more specialised form or function, in essence the acquisition of greater degrees of form. This has until recently, with the successful cloning of Dolly the sheep, been understood as a one-way process. Dolly's cloning was revolutionary because it proceeded by the 'de-differentiation' of cells, so that they could begin the developmental process of differentiation into the various parts of a sheep, once more (Franklin n.d.; Wilmut *et al.* 2000).

The modern construal of biology as subject to processes of forming and unforming, and of form as being a structuring of information, is quite radical and provides direct analogies to Lak cognitive terminology. For what is loss or successful preservation of form and information if not forgetting and remembering?

The feminist critique of such formal and informational ways of envisioning reproduction – that they attempt to dematerialize fertility away from the control of women (who have long been associated with the material, as Judith Butler (1993: 31) writes, 'the classical association of femininity with materiality can be traced to a set of etymologies which link matter with *mater* and *matrix* (or the womb)'; see also discussion in Helmreich 1998: ch. 3) – gives an indication of how suitable they may be for understanding patriarchal Melanesian societies infamous for male appropriations of women's role in childbirth. In fact, sensitivity to the importance of form in Melanesian systems of reproduction represents a significant step forward beyond substance-based approaches. For instance, it provided the key to understanding the supposed 'virgin births' of the Trobriands – where the father makes no substantial contribution to the child, who in that regard is said to derive entirely from the mother's matrilineal *dala* spirit. However, the child is shaped and forming of the maternal substance is due to the father and his *dala*: the child looks like the father, who forms them through a variety of means ranging from massage to gifts of ornaments and decorations, and in so doing separates them from their mother (see Strathern 1988: 236–237).

'Form' actually provides a way to mediate the apparent dichotomy between the physical and the cognitive understandings of biology. 'Form' need not be seen in opposition to content, or material substrate: as a perceptual phenomenon it derives from both subjective attention and objective, physical presence. In fact, the experience of form transcends the otherwise major distinction between mind and matter. We can attempt to inculcate formal qualities, say coherence or clear articulation, in the physical world we engage with, as much as in our thoughts. In Lak, the manipulation of physical forms, such as effigies, are used to effect what might be termed 'cognitive form', for instance by reducing the coherence and articulation of memories.

I argue that to reconcile the cognitive emphasis put forward in Lak, and elsewhere, with substantially construed biological understandings, it is helpful to regard the life cycle as a cycle of form, conceptual form as much as physical. If one considers the beginnings of life in this light, it is generally (always?) supplementary. Whether a chick emerging from an egg, a leaf from a bud or a child from a womb, they each add something to the world. What is added lies on the boundary between the subjective and the objective, the cognitive and the physical, because what is created is form, a new perceptual entity, a new gestalt.

As Barraud *et al.* (1994) suggest for Melanesian contexts, and as Ingold (2002) suggests for a biological discourse not separated from culture, it is probably appropriate to view each stage of the life cycle in Lak as a transformation within a relational field, one of a chain without end with the result of one transformation necessarily being implicated in the next transition. Transformation might be seen as the converse of supplementation, but that is far from the case if one moves from purely material, noumenal accounting to more phenomenological modes, in which transformation is the creation of a 'new' (relatively and cognitively distinct) form from others, one 'added' to the world. It is this kind of morphological, cognitive and physical perpetual redefinition of the entities involved in the life cycle that I now examine.

The life cycle

In Lak, the physical body is inert material, 'wooden' (*yai bobolos* – wood remaining), without spirit, *talngan*. *Talngan* may indeed be defined as 'life', the moment of death is said to occur when one's *talngan*/spirit departs. Life's start, on the other hand, is marked by the arrival of spirit in a child. *-Ngan* is a possessive marker; *tal-* is the root term. *Talngan* is somebody's spirit; *talung* is the generic word for all types of spirits. *Tal* or *taltal* means to wander, or to roam without any specific purpose, which is just how spirits are supposed to act.

Talngan are particularly associated with the visage or countenance. They are likened to the image one sees in the mirror. But when walking in the forest, any sign that reminds one of a dead person — a sound of whistling, or a bird, or a light — may, if one feels it to have the 'mark' of the deceased, be taken to be their *talngan*. In a child it is in their looks and character and actions that their *talngan* will be discerned and identified as that of a deceased person. In many cases children themselves are said to relate whose *talngan* they have and to begin using the kin terms appropriate for that identity.

People are not the only entities with life, with spirit. The land, in particular, is also alive. 'It has eyes, ears and nose,' I was told, 'your land is like your pig; it knows who its owner is.' Many spirits roam the forest, but the most important ones are those which have male initiatory cults centred upon them. There are two main types: one, which could be classified as a 'bullroarer cult', is centred around spirits known as *tamianpopoi* ('the man who eats rotten wood'); and another – the most powerful, prestigious and fundamental to Lak culture – is centred on *tubuan* (locally known as *nantoi*) spirits, which take the form of large, enveloping, forward-inclined masks with prominent eyes. The tubuans are the most powerful spirits, they are violent and homicidal amalgams of the collective dead that take away the final remains of deceased individuals; at the same time they are also mother-spirits, which give birth to child spirits known as *dukduk*. They have a dual and central role in the life cycle, combining the final dispatch of humans with the birth of independent spirits. The also have a powerful and explicit, and correspondingly dual, role in shaping social cognition: on the one hand, their role in mortuary ritual is

Cognitive Aspects of Fertility and Reproduction 165

Figure 7.1. *This mask is a* nantoi, *a mother tubuan, the apical image of the men's initiatory cults and the most fearsome yet seductive sight in Lak*

the removal of the last material that is evocative of deceased persons, so that they may be forgotten; on the other hand, all the living are compelled to gather together and watch *only* the tubuan as it dances and provokes sorrowful recollections of the dead who have been associated with them in the past.

In Lak, life cycle (and 'life'/spirit) transformations are managed largely through intercourse with a spirit world, of which the tubuan is the prime manifestation. The key transformational moments are engineered through a series of rituals, easily divisible into those that concern mortality and those that concern

Figure 7.2. *Dancers with* kabut, *lesser versions of tubuan*

fertility. The former are said to be owned by men, the latter processes belong to women.[3] They are:

1. Men's rites of death:
   (a) primary funerary rites to clear *sum* (negative relations of loss and debt with the deceased) and its restrictions from places and people, culminating in a *tondong* exchange.
   (b) secondary funerary rites, many years later, in which men's initiatory-society tubuan spirit masks take away *nambu* – material evocative of the deceased, including a *lalamar* shell-money effigy.
2. Women's rites of fecundity:
   (a) the *dal* female initiation in which women are made fertile, by spirits.
   (b) giving birth to children, again via interaction with men and spirits.

As in much of Island Melanesia, death garners a great deal of social attention in Lak and is the focal point of politics and ritual. It is death that presents the greatest public spectacle, with the revelation of tubuan spirits that effect the most powerful and overt social control of cognition. Newborn children are also revealed in the form of the dead, but instead of the collective dead (that a tubuan instantiates) they present an individuation and differentiation of ancestral spirit. Correspondingly, birth and its rituals command only a fragment of the attention, and exhibit only a fraction of the aesthetic power of the tubuan, and are of equally reduced social and political import. This is of course not unconnected with the gendered division of the life cycle. The greater elaboration of mortuary rites mean the relationship between the cognitive and formal articulation of the tubuan and the person is more visible than in the reverse processes in rites of fertility, so it is with life's ending that I begin.

Primary mortuary rites

On death the entire cognitive form that the person has built up during their life career is transformed. Those who were incorporated, through the activities glossed as 'thinking of', in the deceased's relational self have lost both someone whom they took care to think of, and someone important in giving them attention. They find themselves afflicted by a loss of social definition. This deficit suffered by people, places and objects associated with the deceased is known as *sum*. *Sum* has two glosses. The first is as a kind of dust or dirt. At death those implicated in the deceased have their visible form diminished to various degrees, from being painted black and remaining in enclosure, to not washing. The hamlet of the deceased is left littered and unbroomed. Objects they looked at before death are broken and scattered. Activities they partook in are neglected by those with who would do them with them. The explanations for these run along the same lines, they not only remind the living of the dead but they are 'dirty' because of the deceased's association with them. For instance, people refuse to eat certain foods, because they remind them of the deceased and how they would 'think of' them with that food, i.e. through sharing or giving. The task of the primary mortuary rites is to think of the village and the mourners anew, and by so doing reform them and replace the attention of the deceased that once formed them with that of the ritual host.

The second sense of *sum* is of a loss, like a debt. A debt is a marker with which one reminds oneself and another that one needs thinking of, that that attention in the form of an exchange good is absent, in abeyance, and morally deserved. This is why, whether or not there were any exchange goods outstanding from the deceased, all the mourners have debt, a debt that has been defaulted. The attention that they were expecting from the deceased has been rescinded, while their thoughts of him linger on.

At the first stage of mortuary process, the initial burial,[4] the two primary activities are the determination of who will take responsibility for ensuing series of rituals and the instantiation of mourning and *sum* in the first sense – the breaking and the dirtying of the objects, persons and places diminished by the death. In particular the spouse and eldest child of the deceased become chief mourners and are coated in a black substance and secluded in the darkness of a house

The following stage dwells more on the second sense of *sum*. This is *anngan* (an imperative meaning 'eat!'), the compensatory feeding of – thinking of – the entire village by local-hosts in all the hamlets where the deceased is known. Those being fed are at least notionally reluctant to take part, and may need coercing to eat. This is partly because the overt rationale of these feasts is that they will think no more of the deceased once the meal is consumed, but partly also because political capital is surrendered to those that feed one. Rather than 'finishing' thought of the deceased, the local-hosts in *anngan* effectively consolidate their village's *sum* debts of thought of the deceased upon themselves. In turn, they too will be compensated and have their thoughts turned from the deceased by a primary host in a further funerary ritual known as *tondong*.

In the home village of the deceased, the leader (*kamgoi*) of their matri-clan holds a series of *anngan*. These are more elaborate than those of the surrounding villages and, together with ancillary rituals, stretch over four days, during which time large numbers of men and women are fed, and the secluded mourners are given staged payments of shell-money to 'clean' them, both literally and in terms of eating, drinking and rejoining village life. During this period the exposed corpse would, in pre-colonial times, have food brought to it also. On the last day the corpse would have begun decomposing and the skull would have been taken and hidden, while the rest of the remains were disposed of in the sea. Just as the body is disposed of, and no longer given food, so the deceased is now no longer to be thought of and the close kin may exit their seclusion.

Cognitive Aspects of Fertility and Reproduction

Transference of pigs\food

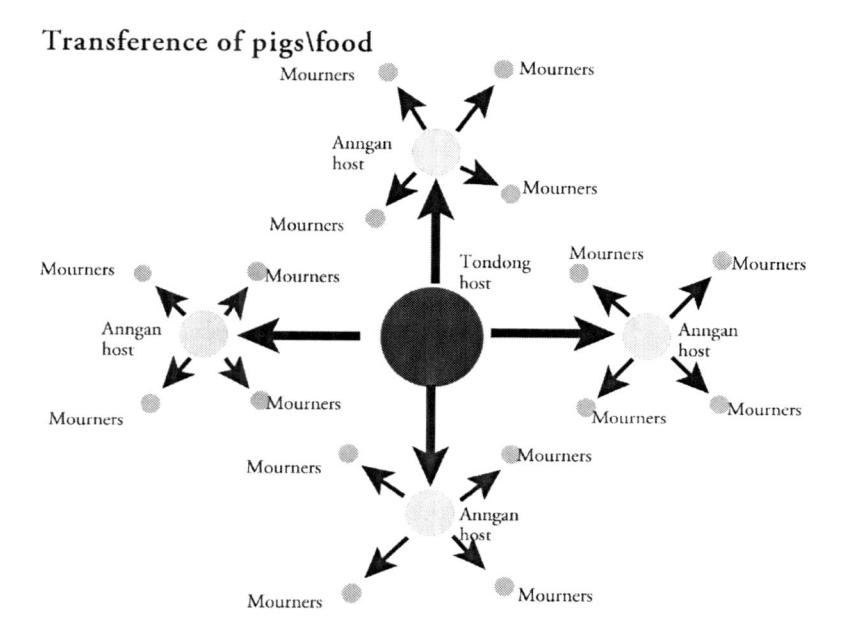

Transference of debt\loss

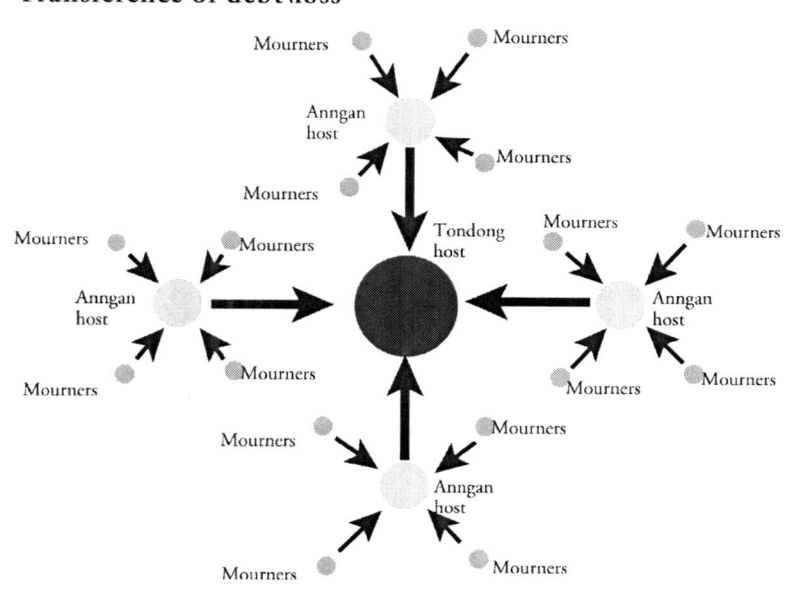

Figure 7.3. *Exchanges at* tondong

Some weeks later the primary rites culminate at a *tondong* ceremony, where the clan-leader brings all the local feast-holders to himself so as to remove their *sum* and further consolidate the debt that absence of the deceased had scattered. Each of the local-hosts brings four things to this central *tondong* at the deceased's village: an effigy called a *tonger*, constructed from food crops; shell-money with which they will contribute to construction of another effigy, called a *lalamar*; a dance troupe; and an exchange pig.

The aim of the *tondong* is spoken of as *tolon ngis*, which literally means to make beautiful or, idiomatically, to 'wash' the family of the deceased and the local hosts. *Ngis*, clean, bright or beautiful is also the word used to describe the decorated men who dance under the auspices of the local hosts. In these dances these hosts and the mourners they have fed are demonstrably relieved from the obscurity of their dirty *sum*. The male dancers come out of seclusion in the bush wearing various spiritual decorations upon their head, known as *kabut*, which are clan valuables of their sponsors and are iconographically and conceptually lesser versions of tubuan. The dancers performatively ensure a redirection of attention, and demonstrate their *ngis* and spiritual state, by approaching the form of the iconic thought-provoking object, the tubuan. Like the tubuan, the dancer's visages are focal points for the remembrance of previous owners, whom the audience may 'see' in their decorations. The men are revealed anew, and made to intimate the tubuan's ideal form by the host's thinking of them. Just as dust, dirt and obscurity of *sum* indicate failure to attend, to know and thence form; so does brilliant decoration and performance indicate and create attention.

The only male dancers who do not wear the spiritual *kabut* on their heads are those dancing under the auspices of the primary host. They dance still smeared in black *sum*, showing that the primary host alone now bears the burden of the deceased. It is he alone that cries at the *tondong* ceremony, it is he that has consolidated the debt that the dead leave behind them by paying and thinking of the living so that they are beautiful and the deceased is thought of no more.

At the end of the day attention turns to the effigies, in front of which the dancing has been taking place. Each is associated with the deceased their social relationships, and must be dismantled by the primary host making payments to the local hosts. The *lalamar* shell-money effigy is most directly said to be the deceased's body and used to be topped by the skull of the departed. The primary host pays each contributor with shell-money of his own, to have them remove their shell-money and disassemble this body once more. Each of the *tonger* food effigies is constructed with produce from the deceased's garden, and it is under these that at the end of the *tondong* the pigs are tied. These are *gar* 'challenge' pigs, and a series of exchanges are initiated in which the host must both give compensation pigs in return for the initial *anngan* expenditure and also replace *gar* in succession with a larger pig of his own. At this point the effigies, which represented the outstanding relationships, thought and debt the local hosts still had with regard to the deceased, are torn down. The rest of the community now have no claim on the deceased, who becomes entirely the host's concern. The host moves from having the responsibility of paying for the debts generated by the death, and begins to give without return, forcing the community to become indebted to him. This is the switching point to the secondary rites in which the host's unilateral giving transforms him into the focus for the entire community.

Secondary mortuary rites

Many years generally pass after the *tondong* before secondary mortuary rites are instigated for the dead. Families and communities have long since accustomed themselves to living without the deceased. Only exemplary rituals hosted by the most powerful men bring forth the tubuan masks, the most powerful and iconic form of spirit, and the incarnations most obscuring and transforming of their bearers. Although hosts may make do with lesser spiritual forms, the role of the tubuan is closely identified with the secondary mortuary rites – as they are the only time they can appear in public.

Once all the tubuan are ready in the *taraiu* – their secret bush realm accessible only to initiates who have 'died' – the host holds a feast at which he feeds, and gives shell-money unilaterally to, the entire populace of the region. This action causes him to be classified as the 'mother' of the community, with power over all that are indebted to his nurture. In fact, the host converts his consolidated dusty *sum* into the most complete rendition of cognitive form, the tubuan. By giving unreciprocated pigs and shell-money to the entire community, giving until he has nothing left, the host becomes himself like a spirit and like a mother who has nurtured and thought of them all and in turn forced them to attend to him. He becomes identified as a mother tubuan.

Some time during the feast, the *mat a matam* 'dead come and look' will occur – tubuan will invade the village from the *taraiu*,

Figure 7.4. Nantoi *tubuan remove 'bones' of the deceased*

to take away the 'bones' of the deceased. These 'bones' are the remaining physical reminders – known as *nambu* – such as the remains of their house, the tree they once sat underneath, or their ceremonial axe. In addition they remove another *lalamar* shell-money effigy, this time constructed by the primary host alone. The women retreat from the masks, which are dangerous to them, watching them from a distance, and wailing as they see those who have died in their faces.

The individual whose death is the occasion of the rite is said to be forgotten from this point on, and is in effect incorporated in the plural lineage evoked by their mask. More than the deceased are elided from social attention during the ritual. Those initiates that are not already hidden within the masks, and new initiates that are seized from their mothers, accompany the tubuan back to the edge of the *taraiu*, where they are symbolically killed by the host's mask before disappearing into the *taraiu*. This is the last time that any of these men may be seen by the women and non-initiates for the duration of the tubuan appearance. The initiates may not even be mentioned and do not 'exist' for the women any more; their identity is completely subsumed in their dancing display of beautiful spirits. Only the tubuan adepts, who are themselves identified with masks and are called by the name of a mask for the duration of the rite, may pass between *taraiu* and village outside of a mask. The host is identified with the premier mother-mask, who has the power of life and death over the entire community. Other initiates have child masks, *dukduk*, without eyes, which mark their subservient position, and show also the reproductive parallels between human and tubuan mothers.

The masks now dance at the edge of the village, morning and afternoon, for four days. The women are compelled to watch these performances, yet at the same time they must avoid the gaze of the masks, lest they too be transformed in a damaging way. While the spirits are abroad, they are under similar conditions to those of *sum* for a recent death.

The tubuan the men incarnate eventually have to die and be forgotten in order that the living men are known once more. After four days the tubuans' eyes are smeared with black *sum*, their dance loses its vitality and they leave to die off-stage, where the processes for their disarticulation and the removal of the *sum* their death produces take place in secret. What is visible to the women is the intermittent reappearance of the initiates as living humans enveloped in decreasing tokens of the tubuan at the edges of the village, before finally reintegrating themselves into social life at the end of a period of *sum*.

Female initiation

The tubuan's overt role is the disarticulation of human gestalts, not their production. But as in many other Austronesian societies, death in Lak is a 'de-conception' (Mosko 1983), a disarticulation of the social relations that the person embodied as well as their physical constitution and indeed their cognitive realization. Conception and birth are the reverse of that process, the physical and social articulation of persons as cognitive gestalts. In Lak, birth and its rites are also construed cognitively, spoken of as a remembering and bringing into mind that contrasts with the forgetting and absenting from mind of death. The 'conception' of babies at their birth only takes place via a disarticulation of the totalizing figure of the tubuan, mirroring the forgetting of the dead as a coherent nexus of social relations and attention by the coalescence of the tubuan form at the other end of the life cycle.

The prior stage to birth in the Lak theory of the body is menstruation, conceived as a rotting initially caused by interaction with spirits and thereafter by intercourse with men. The girls' initiation, called *dal*, is a rite in which first menstruation is provoked and she is made fertile and ready for marriage. The rite has many similarities to those of the tubuan, only some of which I mention here. In particular, the young girl and her enclosure are highly redolent of the tubuan mask. She is painted red, her eyes encir-

Figure 7.5. *A* dal *in her spiritual decoration, note the tubuan roundels on her eyes*

cled with roundels similar to the tubuan's, and she is covered with perfumed herbs both to attract spirits and to hide the smell of her impending decomposition. Her enclosure, within which she must stay until menstruation is deemed to have occurred, is a conical tubuan shaped container within a house.

The hidden space within women's bodies, where menstruation takes place and from whence children are born, is explicitly likened by men to their own secret spaces within masks, which are in turn associated with the dark interiors of houses. The young girl is neither visible nor nameable for the duration of her enclosure, her social form and role neither seen nor enacted. Instead she is treated as a spirit, not allowed to walk on the floor and fed on a diet of greasy substances normally connected with men's preparations for the incarnation of spiritual decorations. Just like tubuan and their owners she is called by a name that is a lineal possession, except this time one that incorporates the identities of a historical sequence of women that have previously been initiated.

Every night while she is enclosed, the women must gather in front of the house to sing *kubak* songs and watch the visitations of the spirits, just as they are obliged to watch the tubuan. These, although encompassing of the men who must not be seen by the women, are free-form and theatrical incarnations of spirits, completely different from the rigid and copyright templates of the tubuan. In one performance these included a giant pig, giant birds and a floating ship. The songs the women sing during the spirits' visitation are of sexual attraction and transparently phallic symbolic accounts of spiritual snakes biting the girl and making her bleed.

When the girl is finally revealed from her tubuan-shaped container, in her blood-like red decoration, she watches a parade of young men in spiritual costumes, this time called *malerra*, the local term for love magic. They try to woo her away from her designated husband – and are far more clearly *men* wearing tubuan inspired designs, than were the spirit forms that appeared during the nights of her seclusion. Ideally the final act of her initiation is marriage on a subsequent day to her fully human, non-decorated husband.

During *dal* the girl is linked to tubuan in an over-determined fashion and her decomposition is firmly associated with the hidden and internal while being correlated with the decreasingly spiritual and increasingly human nature of her male suitors. Whether or not she has physically menstruated, her spiritual interior is imaginatively and cognitively decomposed in an analogous manner to that in which the social person is decomposed in mortuary rites. In the *dal* ritual the form of spiritual performances are

Figure 7.6. *A less formed male spirit (*tamsaikio*) appears at night*

manipulated to shape the cognitive form of the girl's social body. In mortuary ritual it is exchange-good effigies that are disarticulated to decompose the social person.

Birth

Conception beliefs are that the child is further coagulated and dried male blood. The baby is dried into a firm and formed being, both inside the mother and by the application of hot leaves after birth. The mother's role in birth is to feed and grow the child both within her body and when it is born. She feeds it with a variety of liquids — breast milk, coconut water, sweated greens — known as *polonon*, the term for the decomposition fluids from human or animal bodies. In fact there are a number of myths of children prospering by eating the decomposition fluids of their dead mother, and it is not too great a condensation to see these dead mothers as *the* dead

Figure 7.7. Malerra, *men with various owned spiritual decorations and qualities*

mother, the tubuan. This reinforces the thesis that the child's composition is at the expense of the tubuan's decomposition.

The mother gives birth in seclusion in a house, accompanied only by mothers already 'initiated' by child-birth – a restriction overtly likened to that of the seclusion of male initiates with the tubuan. The mother and child should stay hidden together for four days after birth – the same time required for the initial disarticulation of the deceased on death. When the child is revealed from the obscured spiritual interior of a woman and a house, its dark pigment has not spread and he or she is white. In this state the child is seen as analogous to the white-shell-money *lalamar* effigy taken into obscurity by the tubuan. *Lalamar* are not only used in mortuary rituals, but are also used in healing practices as substitutes for sick children, into which an illness-causing spirit may be lured. Like the *lalamar*, the child is a body that hosts a spirit. Their spirit, *talngan*, is the life and consciousness of the child, and was previously another person. It is expected that the particular deceased will be recognized from the face of the child, or that the child himself will announce or exhibit who they are. This attribution creates bonds between the child and the family and friends of the dead person.

Thus while still young, the *form* of a child, their very appearance and behaviour, draw attention to who she or he is. People do not know who a child really is until they recognize them, until they are moved to see it in their visage. Is what they see then a memory? In one sense this is form as memory and indeed social relations – all three discerned as one. But the child does not merely evoke a dead spirit, he or she is that spirit; not a memory *of* a deceased, but the *same* deceased. As de Coppet (1981) notes, persons are reformed in many different ways in the course of the transformational chains of ritual, yet still the cycle reproduces the 'same thing'. Memories are virtual, merely derivatives of the past. The child is instead a refocusing of attention in the present tense. The child is a re-articulation of someone whose social form has previously been disarticulated through the redirection of social attention upon the tubuan. Their individual identity is only possible by the de-totalization and fragmentation of the tubuan's collective identity as demonstrated through the spiritual decomposition of *dal* and birth.

This is, of course, only the beginning of the process. Once revealed from the inchoate and obscured spiritual zones within women, the child can be seen to become an object of knowledge, a more determinate form. That form becomes more determined and sharply defined through processes of attention. He or she is thought of through gifts of food and through recognition as a certain person. As children become adults and proceed through their life-course they continue to form themselves and others through the action of reciprocal attention. Without thinking of each other people would have no form, would be unknown. This social attention is at the very heart of Lak sociality: it is the means by which people know each other and themselves.

Conclusion

This chapter, through examination of a case-study from Lak, attempts to demonstrate that the 'biological' realms of reproduction and fertility have a cognitive dimension. Though this may seem foreign to Western, modernist intuitions, this cognitive perspective is made more commensurate with 'materialist' understandings through consideration of the mediating quality of form. Form, or even 'organization', as a perceptual phenomena, can readily be seen to be as a juncture of both material and cognitive realms. 'Reproduction', whether abstractly or experientially considered, is a process with a strong formal element to it: some additional form is (re)produced. That process has dual effects, cognitive and material. Conversely, the ethnography in this chapter shows how dual modes of action, material and cognitive, can effect reproductive processes, because, of course, reproduction *itself* is a juncture of those realms.

Acknowledgements

I thank the people of Lak, in particular those of Siar village, for their hospitality, kindness and patience during my fieldwork of 1994–1996, which was undertaken during my period as a postgraduate student at University College London and was funded by the ESRC. I am grateful to Soraya Tremayne and Stanley Ulijaszek for an invitation to present an early version of this article at the Fertility and Reproduction Seminar of the Institute of Social and Cultural Anthropology, Oxford.

Notes

1. Lak is one of New Ireland's largest (just over 1,000 km²) and least populated (a little over 2,000 people at the time of my fieldwork) administrative areas, and comprises the eastern half of the mountainous 'bulge' at the southern end of the island. In 1996 most of the population relied on subsistence gardening and lived in small villages and hamlets along the narrow coastal strip. Within New Ireland they have a reputation, perhaps because of their relative remoteness from towns, for the strength of their *kastom* ('traditional' beliefs and practices).
2. Of course, notions of the embodied mind have been in vogue within academia for some time (e.g. Johnson 1987, Edelman 1992), but these mainly come from students of the mind rather than the body, and serve more to give a physical cast to the mind than a cognitive cast to the physical.
3. The following account of the life cycle is similar to that published in Kingston (2003).
4. A Christian innovation; pre-colonially the body of the deceased was exposed.

References

Allen, M.R. 1981. 'Rethinking Old Problems: Matriliny, Secret Societies and Political Evolution'. In *Vanuatu: Politics, Economics and Ritual in Island Melanesia*, ed. M. Allen. London: Academic Press.

Barraud, C., D. de Coppet, A. Iteanu and R. Jamous 1994. *Of Relations and the Dead: Four Societies Viewed from the Angle of their Exchange* (trans. S. Suffern). Oxford: Berg.

Battaglia, D. 1990. *On the Bones of the Serpent: Person, Mortality and Memory in Sabarl Island Society*. Chicago: Chicago University Press.

_____. 1992. 'The Body in the Gift: Memory and Forgetting in Sabarl Mortuary Exchange', *American Ethnologist* 19: 3–18.

Butler, J. 1993. *Bodies that Matter: On the Discursive Limits of 'Sex'*. London: Routledge.

Coppet, D. de 1981. The Life-giving Death. In *Mortality and Immortality: the Anthropology and Archaeology of Death*, eds S. Humphrey and H. King, 175–204. London: Academic Press.

Edelman, G. 1992. B*right Air, Brilliant Fire: On the Matter of the Mind*. New York: Basic Books.

Franklin, S. n.d. 'Biological Propriety', paper prepared for 'Forms of Intellectual Creativity' panel at the Property, Creations and Transactions conference, Cambridge University, 13–15 December 2001.

Helmreich, S. 1998. *Silicon Second Nature: Culturing Artificial Life in a Digital World*. Berkeley: California University Press.

Ingold, T. 2002. 'Between Evolution and History: Biology, Culture, and the Myth of Human Origins', *Proceedings of the British Acadamy* 112: 43–66.

Johnson, M. 1987. *The Body in the Mind: the Bodily Basis of Memory, Imagination and Reason*. Chicago: University of Chicago Press.

Kingston, S. 1998. Focal Images, Transformed Memories: the Poetics of Life and Death in Siar, New Ireland, Papua New Guinea. Ph.D. thesis, University of London.

_____. 2003. 'Form, Attention and a New Ireland Life Cycle', *Journal of the Royal Anthropological Institute (N.S.)* 9: 681–708.

Knauft, B. 1989. 'Bodily Images in Melanesia: Cultural Substances and Natural Metaphors'. In *Zone: Fragments for a History of the Human Body, Part 3*, eds M. Feher, R. Nadaff and N. Tazi. New York: Urzone.

Küchler, S. 1988. 'Malanggan: Objects, Sacrifice and the Production of Memory', *American Ethnologist* 15(4): 625–637.

_____. 2002. *Malanggan: Art, Memory and Sacrifice*. Oxford: Berg.

Lutkehaus, N.C. and P.B. Roscoe, eds 1995. *Gender Rituals: Female Initiation in Melanesia*. London: Routledge.

Maschio, T. 1994. *To Remember the Faces of the Dead*. Madison: Wisconsin University Press.

Mosko, M. 1983. 'Conception, De-conception and Social Structure in Bush Mekeo Culture', *Mankind* 14(1): 23–32.

_____. 1989. 'The Developmental Cycle Among Public Groups', *Man (N.S.)* 24: 470–484.

Strathern, M. 1988. *The Gender of the Gift*. Berkeley: University of California Press.

Thomas, P. 1999. 'No Substance, No Kinship: Procreation, Performativity and Temanambondro Parent-child Relations'.

In *Conceiving Persons: Ethnographies of Procreation, Fertility and Growth*, eds P. Loizos and P. Heady. London: Athlone Press.

Thompson, D.W. 1961. *On Growth and Form*, abridged edn, ed. J.T. Bonner. Cambridge: Cambridge University Press.

Wilmut, I., K. Campbell and C. Tudge 2000. *The Second Creation: The Age of Biological Control by the Scientists who Created Dolly*. London: Hodder Headline.

CHAPTER 8

HISTORY EMBODIED: AUTHENTICATING THE PAST IN THE NEW GUINEA HIGHLANDS

Michael O'Hanlon

I begin with two words, former denizens of every seminar but which have since been exorcised so totally from most anthropological discourse that they might never have existed. The two words are: Levi-Strauss.

For anyone interested in the relationship between artefacts and history – whether or not in the context of thinking about fertility and reproduction – Levi-Strauss's (1966: 238ff) reflections on the Aboriginal ceremonial boards known as *churinga* are characteristically exhilarating. The passage in question follows Levi-Strauss's contrast between 'hot' and 'cold' societies: between those (as in the West) that internalise history and make it central to their workings and those (such as Aboriginal Australia) which seek to 'annul the possible effects of historical factors on their equilibrium and continuity ...' (1966: 234). Levi-Strauss goes through the various reasons that have been advanced for the sacred status of *churinga*, discounting each, before suggesting that they owe their status not to their content but exclusively to their form. In the otherwise synchronic Aboriginal world, which at one level collapses the distinction between ancestral beings and living men, *churinga* are the past materially present. They have a 'probative' function, in that they alone testify to diachrony in an otherwise synchronous world.

What Levi-Strauss (1966: 238–239) further suggests is that when an exotic custom exercises a powerful fascination on us – as *churinga* undeniably have done – 'it is generally because it presents us with a distorted reflection of a familiar image, which we confusedly recognize as such yet without managing to identify it'. Here, the 'familiar image' is that of the originals of our own documentary archives. Like *churinga*, he argues, documentary archives are *not* valued on account of what they tell us about the past which is independently attested through copies, through books, and through commentaries. Rather, what both archives and *churinga* do is to put us in touch with 'pure historicity' (Levi-Strauss 1966: 242). It is for this reason, not on account of their content, that both are treasured, he argues.

I will return to Levi-Strauss in a moment. My purpose in this chapter is to outline the dynamics of a knowledge system – one integrally connected to issues of fertility and reproduction – from the Wahgi people of Papua New Guinea's Western Highlands Province. In regional terms, this knowledge system exhibits a number of points of interest. Papua New Guinea (PNG) is noted for cultures with cosmologies of great intricacy, whose interpretation has been the focus of discipline-wide debate. One thinks here, for example, of Barth's (1975, 1987) accounts of the Ok initiation systems or of the debate over 'misconstrued order' in Melanesian religions (Brunton 1980a), which was partly prompted by Alfred Gell's (1975) volume *Metamorphosis of the Cassowaries* which itself precipitated an edited collection entirely given over to re-thinking the interpretation of this single West Sepik ritual (Juillerat 1992; see also Brunton 1980b, Juillerat 1980, Gell 1980 and Whitehouse 2000). But most such ethnography has emanated almost exclusively from lowland and fringe areas, and not – as in the present case – from the central Highlands which have tended to be categorised as robustly 'thing' rather than 'knowledge' oriented (Errington 1988: 766).

A second point of interest is that while the Wahgi knowledge system on which I shall focus is concerned with the same themes of strength, fertility and reproduction as are the focus of Ok and Sepik cosmologies, the knowledge at issue in the Wahgi case hinges on inter-personal issues, and relates to sorcery, to witchcraft and to betrayal.

A third point of interest from the specific Wahgi material lies in its concern with how knowledge is authenticated, how Wahgi know things to be true. This is an aspect that is largely absent from the analyses of other Melanesian cosmologies, where the focus has tended to be more on how anthropologists know things to be true: for example, whether ritual is better understood in terms of digital or analogical models (Barth 1975: 207ff), or whether the academic privileging of coherence has led anthropologists to assume that indigenous cosmologies possess greater logical coherence than in fact they do (Brunton 1980a, b).

Returning now to Levi-Strauss's reflections on the role played by artefacts in authenticating history, my overall argument will be that the Wahgi do indeed seek a form of external verification in this context, but with some major qualifications. The first is that the probative form to which the Wahgi turn lacks the apparently enduring nature of *churinga* or of safeguarded documentary archives. Here I will show how this role may be taken by the transient qualities of the human body itself, and its entailments in the form of 'skin', performance, and patterns of fertility. My more major point, however, is that the Wahgi problematic is a very different one from that delineated by Levi-Strauss. Rather than needing probative keepsakes to attest to diachrony, for the Wahgi, history is axiomatic. This history takes the form of hundreds of accounts detailing relationships and actions, some in the remote past, some more or less contemporary, some covert, others not. What they have in common is that they all describe events and actions which are thought to continue to influence, and at times to determine, the shape of the present. The Wahgi, then, see themselves as inextricably embroiled in history – as in their own persons the outcome of it. Contra Levi-Strauss, the Wahgi problem is not to be reassured of diachrony: their dilemma rather is to distinguish the authentic from the inauthentic from among the plethora of rival histories offered them. And in attempting to do this they turn to the human body, its state and condition, the quality of the performances in which it is engaged: and to its fertility. What I shall try to show is how this whole range is constantly scanned by the Wahgi as potentially bearing witness to the otherwise hidden state of crucial moral relationships whose antecedents are located in the distant past yet continue to stamp themselves upon the present.[1]

Intersecting principles

Amongst the most effective accounts of Melanesian societies have been those in which the complexity and tenor of social life is traced to the dynamic interplay of a pair of opposed principles. Sometimes the principles in question have been broadly phrased as values: for example, the interplay of 'strength' and 'equivalence' which Read (1959) long ago identified among the Gahuku Gama. At other times, the principles have been more specifically social structural: for example, the interaction of 'consanguinity' and 'exchange' among the Daribi (Wagner 1967), the contradiction between 'siblingship' and 'descent' in Etoro (Kelly 1977) or, more recently, that between totemic clanship and community identified by Harrison (1993) as underlying Avatip life.

Much of the social life of the mid-Wahgi people of the Western Highlands Province of PNG can similarly be encapsulated. In the mid-Wahgi, my wife Linda Frankland and I have worked principally with Komblo, one of the paired tribes living in the northwest part of the area, only a few miles east of the linguistic boundary with Hagen peoples. Like Hageners, the mid-Wahgi's first exposure to the outside world came in April 1933 with the celebrated Leahy-Taylor patrol (Connolly and Anderson 1987). Like Hageners, the mid-Wahgi subsequently underwent a crash-course in modernity, successively being caught up in the periphery of the Second World War, acquiring Australian and subsequently PNG currency, taking to cash-cropping coffee, experiencing the inflation of ceremonial payments brought by the rise in coffee prices; and later suffering, in places, the return of inter-group warfare, made worse since the 1980s by the introduction of firearms (O'Hanlon 1993).

In the Wahgi case, the principles to whose interplay the complexity of traditional Wahgi life can be traced are the rival claims of clanship and of what Wahgi term their 'source' people – especially but not exclusively their maternal kin. It is necessary to provide a thumbnail sketch[2] of each because it is Wahgi narratives about these two sets of relationships which make up the kind of history in which they are most interested, and which in turn bear closely on issues of fertility and reproduction.

At one level, there is great emphasis in Wahgi life on the strength, the solidarity and the capacity to project itself of the group, especially of the exogamous patriclan which, locally, constitutes the basic political unit. Traditionally, this emphasis was expressed most publicly in the performance of the once-generational Pig Festival (Figure 8.1). This is put on autonomously by the clan, and involves clansmen ritually segregating themselves in particular ways and then presenting themselves as dancers to the scrutiny of allies, rivals and enemies (O'Hanlon 1989, chapter 4). At the climax of the Pig Festival, the majority of a clan's pigs are clubbed around a ritual structure whose timbers have been secretly excavated from the spot where they were buried at the close of the clan's previous Pig Festival a generation before. This

Figure 8.1. *Yimba! Aipe (later member of parliament for North Wahgi) dancing in the Pig Festival, 1979. A compelling appearance, along with the birth of children, is taken to testify to moral probity.* © Michael O'Hanlon.

ritual structure is in one sense a specific model of the enduring clan, for each of its supporting timbers may be said to stand for one of the constituent subclans; moreover, a peculiar property of the timbers is that they are said to endure, as the clan is supposed to, and never to rot, however long they remain buried. It is thought that if clanspeople are united, work hard, and observe the relevant ritual restrictions, the Pig Festival will ensure both the future fertility of the clan's pigs and of its members, as well as securing the occupancy of its land in the face of rivals and enemies. In recent years, full-scale Pig Festivals have been uncommon in the mid-Wahgi area, but many of their values are still encapsulated in the accelerated exchange events called in Tok Pisin *pati* ('parties').

But if at one level health, strength and prosperity are traced to clan cohesiveness and unity as manifested in the Pig Festival, at another level they are traced to an individual's 'source people', known as *pul alamb* or *mambnem*. Maternal kin are the primary example of 'source people', and ego's maternal kin are owed 'strength' payments throughout life, and also at death. In return, maternal kin have the capacity to bestow health and well-being on their sisters' children or, if angered, to curse them, resulting in infertility, illness and death. However, the category of 'source people' is wider than maternal kin alone. It also includes members of an individual's mother's father's clan, mother's mother's clan, and – insofar as they are known – members of mother's mother's mother's and mother's mother's father's clans: in short, any group which has in the past provided a woman from whom ego is descended. There are also surrogate 'source people' within ego's own clan, comprising those who were the 'road' along which ego's mother came in marriage to ego's father, or the descendants of such people. Wahgi think of any new marriage as a difficult enterprise, requiring the aid of a 'road link person', whose ownership of a pre-existing marriage 'road' must be acknowledged and rewarded. The major 'road' for any new marriage is generally the last woman to have married into the clan from the same external group now providing the new bride, or the male children of such a predecessor. Such 'road link people' are explicitly likened to maternal kin, with the difference that they are *within* the clan: like maternal kin they 'brought' a woman in marriage, and they possess all the powers to bless and to curse attributed to maternal kin, whose life-cycle payments they may also receive if relations with real maternal kin are disconnected through warfare or for other reasons.

In Wahgi thinking, intermarried groups should not make war on each other. There should, therefore, be no contradiction between maintaining good relations both with clansmen and with 'source' people. In practice, however, intermarried groups have regularly fallen out and fought, potentially engendering terrible dilemmas for clansmen who find themselves ranged against a group including their maternal kin or their children's maternal kin, who have all the power to bless or to withhold blessing. Now, Wahgi marriage patterns are relatively restricted, with a considerable amount of repeated intermarriage between the same clans over the generations. Every clan is therefore likely to include what, borrowing a term from Meigs (1984: 13), I have elsewhere referred to as 'shadow communities': that is to say a number of blocks each of whom identify a different external clan as their 'source', and to whom they owe their allegiance. These 'shadow communities' can at times emerge as a major threat to the unity of the clan within which they are embedded. Men will refuse fully to support their fellow clansmen in warfare where their opponents include 'source people', and may be suspected of revealing battle plans to their source people. Again, warfare between two other clans into which a given clan is heavily intermarried may cause that clan to split (*bou nim*) as its clansmen rush off to support their 'source people' on either side, thus ending up on opposite sides of the battlefield to their own clansmen.

In passing, it might be suggested that this external, rival, source to which Wahgi attribute health, strength and fertility, is actually tacitly acknowledged at the heart of the Pig Festival, that celebration of the clan as the author of its own well-being. This lies in the fairly open secret that the ritual structure erected at the climax to the Pig Festival, and modelling the clan in its own construction, is not built out of the *mond* wood in whose leaves it is decorated, but out of a different kind of wood altogether: paralleling the point that the clan, too, is not the purely agnatic creation which the Pig Festival otherwise presents it as being.

Treacherous clansmen and affronted 'source people'

What I have sketched so far are two sets of sometimes conflicting relationships – those within a clan, and those an individual has with his or her 'source people' – both of which are felt to govern health, strength and well-being. However, these boons are also considered to be imperilled by tensions characteristic to each of the two sets of relationship, and it is these and their particular bearing on issues of fertility and reproduction that I now need to address.

Despite – or perhaps because of – the great emphasis on clan cohesiveness and solidarity, there is a pervasive fear that the clan harbours traitors: disaffected individuals prepared secretly to betray their fellows. *Kum* is the term for such actions, which include witchcraft but also poisoning and a variety of sorcery known as *simbem ngo*: 'giving legs'. While witchcraft is believed to be carried out by both men and women, poisoning and 'giving legs' are generally male arts and, crucially, are thought to involve co-operation with enemy clans. It is enemy groups which are thought to supply poison, while the sorcery practice of 'giving legs' entails a disaffected clansman secretly passing some item belonging to the betrayed individual to the enemy, which is thought to bring about the betrayed man's death (O'Hanlon 1989, chapter 3).

Among the reasons advanced for such treachery are fairly conventional ones: grudges, jealousy, anger at adultery or the disposal of women in marriage, and a wish to ingratiate oneself with the enemy. There are, however, a number of interesting and pertinent aspects to this cultural syndrome. The first has to do with the ramifying consequences with which betrayal is credited. Not only is the bewitched, poisoned or otherwise betrayed individual thought likely to die, but so long as the perpetrator's actions remain secret, the clan more widely is felt liable to be afflicted with infertility, misfortune, defeat in battle and unimpressive appearance. So narratives of betrayal account not simply for the death of one individual, but link and explain as a logical cascade a whole series of misfortunes. The motivating force here is generally felt to be the ghost of the betrayed individual who visits these misfortunes on his surviving clanspeople for continuing to associate, albeit unknowingly, with his murderer. To adapt a term from Whitehouse (2000: 58), there is a potent 'implicational logic' at work. For example, the fact that ego's father died young, that to date his wife has conceived only daughters, and that most recently his own 'skin' and that of his brothers looked 'ashy' while they danced in the Pig Festival, may all be linked and explained in terms of an originating betrayal two generations back in which – unbeknownst to ego – his father's father was poisoned by his own brother with whose descendants ego has unknowingly continued to associate.

A second pertinent aspect to this cultural schema is its crediting of knowledge with inherent potency, and a consequent elaboration of ideas about its concealment and revelation. It is thought that the mere knowledge that covert treachery has occurred within another clan can be utilised to defeat that clan in warfare. Such knowledge is carefully husbanded as a military good, reserved for deployment at a time of maximum advantage. Weaponry itself is regarded as having the capacity to sniff out disunity among opponents and to kill those involved. This applies not merely to the traditional armoury of spears, bows and arrows but to their modern replacements: 'the foreign guns have eyes too', as the Wahgi scholar John Muke (1993: 262) was told in reference to the rash of battle deaths among one Wahgi group, the Kondika, whose Konumbka opponents harboured knowledge of Kondika betrayals.

Equally, there is a cultural elaboration on ideas of confession, or revelation of such knowledge, something that is often part of peace-making with erstwhile enemies. It is only when knowledge of a past intra-clan betrayal is made known (either through voluntary renunciation at a peace-making ceremony, or perhaps through a deathbed confession, or attendant on conversion to Christianity) that the cascading chain of misfortunes can be terminated. Traditionally, this often involved the victim's kin establishing a fire-taboo and ceasing to eat with the traitor and his kin; alternatively, the victim's kin might simply move away and take up residence elsewhere, perhaps with maternal kin or other 'source people'.

Just as divisions are feared to exist within the ideally united clan, so too the relationship of reciprocal nurture and acknowledgement which should exist with 'source people' may equally bear the imprint of a hidden past. Once again, the roots are often traced back to episodes in inter-group warfare. Persisting misfortune – infertility, illness, poor appearance or death in successive generations – may be revealed as having been caused by a man's father having unknowingly slain someone from his mother's brother's subclan in some long past conflict. Conversely, I was told: *Maiam pen to nonange ngal no morange*: 'had you licked blood [i.e. slain affinal or maternal kin] there would be no children'. Alternatively, persisting misfortune may be traced to the fact that, unknown to him, ego's own subclansmen are secreting knowledge about betrayals within ego's wife's clan. Holding such knowledge against a group who have provided you with a wife is a cause to which infertility is regularly traced.

But whether it is to relations among agnatic kin or with 'source people' to which misfortune is traced, demography (along with appearance) emerges as a highly charged domain in Wahgi society, one pregnant (one might say) with moral implications. Not realising this, my wife and I were often mildly surprised in the early days of our fieldwork at the pointed manner in which people furnished genealogical details: 'His first wife had no children, his second wife only a daughter!'. 'All four brothers died!'. 'After he moved to live here with us [i.e. implicitly away from treacherous clansmen] his wife had 3 sons, all fine young men!'. Initially we did not realise that in constructing a web of triangles and circles we were also charting a moral and historical landscape.

In dealing with generalities, I have probably failed to convey the flavour of the kind of narratives I have been describing, the way in which they furnish a compelling framework in terms of which to understand as a comprehensible series otherwise random misfortunes. Let me conclude this section with a single example which entails both intra-clan betrayal and the fear of offending 'source people', and provides an explanation for what is seen as an ensuing pattern of mortality and fertility.

The narrative in question is one of many which surrounds the deaths 60 years ago of Dop and the big-man Ding, both of the Omngarkanem subclan of Kulkanem clan.[3] The originating act for this trail of disasters was Dop and Ding arranging a marriage for one of their subclansmen. This is said to have infuriated another of their subclansmen, Du, who felt he had prior rights to the bride in question. Together with other conspirators in Omngarkanem subclan, and within the wider clan, Kulkanem, Du resolved to kill Dop and Ding. He is said to have secretly obtained some of the feared poison *enz kongo* (literally 'intestine ginger' or 'shit ginger') from the enemy tribe of Kulaka, who also gave Du some shell valuables as an inducement to carry out the act. The shell valuables were shared among the conspirators. Dop is said to have been the first victim, after some of the poison had been inserted in some fruit pandanus he was consuming. Dop's intestines are said to have 'burst' and he was carried dying into a men's house where he asked for an axe and chopped at the main house post, indicating his intention to return as a ghost to kill his treacherous clansmen. Ding was the next victim, after the conspirators had slipped more poison into bananas at Dop's funeral. One of the conspirators is also said to have revealed his treacherous actions to men in Aikup clan, this being information of which Aikup could make magical use in warfare against Omngarkanem.

But now the poisoners are said to have started to reap the whirlwind at the hands of the vengeful ghosts of those they had poisoned. Du himself died, as did the conspirators, one after another. Moreover, an Aikup recipient of Du's treacherous confidences himself revealed what Du had done, since he had an Omngarkanem wife and feared that his children would die if he held such information against his offspring's 'source people'. Nevertheless despite this revelation and a subsequent confession, half a century later people still considered that further details remained hidden. It was conjectured that a recent Aikup bride marrying into Kulkanem would be unlikely to have children; similarly the fact that the son of one of the victims had a batch of male children was attributed to the fact that he lived separately from his subclan brothers, safely apart from those involved in the cauldron of internecine murder and ghostly retribution.

From genealogical and other evidence, the deaths of Dop and Ding, and those others that are said swiftly to have followed, took place in the early 1940s. These particular deaths were, in fact, almost certainly the result of an epidemic of bacillary dysentery which spread through much of the Highlands, including the mid-Wahgi area, after its introduction by U.S. military personnel in August 1943. As John Burton (1983) who has studied the epidemic shows, it caused many hundreds – possibly thousands – of deaths: perhaps in part because of traditional Highlands mourning practices. These bring many people together in the context of cooking and sharing pork, and would have served as a potent vehicle for the transmission of dysentery. One specifically Wahgi cultural practice, since defunct, may have contributed to making the epidemic unusually severe in their particular case. This is the ritual known as *ngumb nzengimb kong*: the 'nose dirt pig'. After a suspicious death, as with those of Dop and Ding, the knife used to cut up the funeral pigs is first smeared with grime scraped from the surface of the corpse's nose. Anyone implicated in the death would, it was thought, choke on the pork and die that night or shortly afterwards. Here then there is a terrible coincidence between the objective factors favourable to transmitting dysentery, and Wahgi understandings which see individual misfortunes as part of an interlinked series reflecting betrayal and ghostly retribution. Whether thinking in Wahgi terms, or in biomedical terms, each funeral can plausibly be seen to set the next in train. Burton (1983: 260) observes that severe though the epidemic was, it did not cause the social collapse which epidemics brought about among newly contacted peoples elsewhere. So far as the mortality in itself is concerned, Burton is probably quite correct. But it would not do to underestimate the dramatic and continuing impact that long past epidemics such as these may have among a people whose cosmology drives them to make sense of individual births and deaths as an interconnected series, reflecting hidden intra-clan treachery or breaches of proper relations with 'source people'.

Indeterminate narratives

However, for all the importance with which such narratives of betrayal and relations with 'source people' are credited, there is also a fundamental uncertainty attached to them, as I said at the outset. There are good reasons why this should be so. The first is that the narratives, as we have seen, are definitively secret: they describe actions that are carried out in secret, they are themselves secreted as a form of military intelligence. Then, even when such narratives are revealed or confessed, it is done in circumstances which do not necessarily favour consensus: for example, a deathbed confession to which only a limited number of people are directly privy.

Let me take as an example two of the many rival versions, recorded in the 1980s, which have as their core why two women, Waiang and Ner, natally from Waplka clan but married into Komblo Kekanem clan, either have no children or had children who died in infancy.

The first version traced these misfortunes back to the 1930s, or earlier, long before the relevant marriages took place, when Komblo were at odds with Waplka. At that time, according to this version, a number of Komblo, including men from Kekanem clan, secretly killed the Waplka man Monye. Twenty years later, two Komblo Kekanem men took Waiang and Ner in marriage, but the wives failed to bear children, or their children died young. Moreover, the son of another Waplka girl married into Kekanem had skin that was lustreless and appeared 'dirty' (*Nganz nzengimb tom. Pende talang no-dom*). It was only when accusations of witchcraft started to be made that an elderly Kekanem man spoke up, and revealed how Monye had been killed. Monye would have been the classificatory mother's brother of the children of his 'sisters', Waiang and Ner, and thus in a position to kill them, to mark his displeasure at their marrying into his killers' group.

The second version also traces Waiang's infertility and the deaths in infancy of Ner's children back to even earlier warfare and killings. In this alternative account, Waplka men had been bribed to take part in a plot to kill two prominent Komblo Kekanem men. The plotters shared out a very long string of cowrie shells, to which a woman's apron string was also attached as an indication that a girl was part of the bribe, and the two Kekanem men were duly killed. Waplka men themselves were not directly involved in the killing but they had handled the cowrie string. Unknowingly, the Waplka men's grand-daughters, Waiang and Ner, married in to Komblo Kekanem, but the fact that they were descendants of the Waplka plotters meant that they had no children, or that their children died. Moreover, many offspring of Komblo women married into Waplka also died, cursed by the fact that they had touched the cowrie shell inducement to kill their own maternal kin.

What is important to emphasise here is that these are not rival explanations held by different people. Both explanations were acceded to at separate times by the *same* man, the woman Ner's husband. These narratives are 'indeterminate' in the sense that people have in their minds a number of possible explanations for patterns of infertility, death, illness and poor appearance, favouring now one, now another, as unrolling events and disclosures dictate, while still retaining a suspicion that all are tangential, and that the real truth is still to emerge.

To cast this process in slightly more abstract form, one might adapt the terms which Frederik Barth proposed as a means of understanding innovation in Ok cosmology and ritual performance. Barth is concerned to understand how the relatively small Ok area comes to possess the great variety of ritual traditions that it does. Here he works from the fact that many Ok rituals (often connected with initiation) are performed only many years apart, and rely on the memory of ritual experts to recreate them each time. To account for incremental change, Barth (1987: 29ff) proposes a model of what he terms 'subjectification' and 're-objectification'. He suggests that innovation takes place during the long period between performances, when knowledge of the ritual in question is held only in the minds of a few ritual experts who creatively modify it according to their individual psychological preoccupations, before re-externalising it in the next performance. A similar oscillation between 'subjectification' and 're-objectification' takes place in the Wahgi case. As I have suggested, an individual will have in mind a number of narratives each of which potentially explains as a linked series a variety of misfortunes. This corresponds to Barth's period of 'subjectification', during which the individual in question will mentally find support for one or other narratives as social life unrolls but may not feel certain enough to voice any of them very publicly. But, in due course, what seems to the individual a compelling concatenation of circumstances may arise. This may be enough to prompt him or her to privilege one of the possible ways of interrelating the misfortunes and, in a moment of 're-objectification', to externalise it in the form of a public explanation, an accusation or a confession.

Analytical language aside, the indeterminacy of narratives lends to Wahgi life a considerable nerviness (see Bercovitch 1989: 139 for a parallel situation among the Nalumin). Beneath what is often a robustly cheery exterior Wahgi are wary: entirely unknown to you, your next door neighbour may have betrayed you; or unknown to him, his grandfather may have been implicated in killing your own mother's brother in some distant conflict. Could it be this that explains your child's sickliness, and your brother's early death? Or something else? One result is considerable residential instability, with people making frequent experimental moves, to see if ceasing to associate with this neighbour, or that clansman, brings to an end what is seen as a chain of misfortunes.

Are there no means of resolving uncertainty? There is certainly a wide variety of means of divination to determine the causes of misfortune but while these might specify what was causing the misfortune, and which pigs might be killed to alleviate it, there seems to have been a degree of reticence in naming the guilty parties themselves; or, when the divinatory process did identify the guilty parties, it was carried out in secret and did not necessarily lead to widespread consensus. Or, as in the case of the 'nose dirt pig' divinatory ritual described earlier, identification and punishment of the guilty are placed in the hands of the dead.

The significance of objective signs and omens

In the absence of certainty, Wahgi scrutinize social life intently for clues and signs, known as *jep*. These signs are in one sense brief eruptions into the overt world from the hidden realm – the location in Wahgi thinking of all those powerful actions and events which actually determine what happens in social life (see also Bercovitch 1998: 217). Such signs, the Wahgi consider, may be made both by the living and by the dead. People are always on the look out for irregularities in the behaviour, dress, or actions of others, in case the person in question is covertly signifying that something is amiss, or that they possess some knowledge bearing on others present. Someone who mis-performs an action – stumbles more than once on a walk, or fumbles in doing something – may be asked humorously but with an underlying nervousness: *Nim wei enen mo jep enen?*: 'Are you doing that for real or trying to tell us something?'. Equally, such signs may come from the dead: a bark belt that slips unexpectedly down, a feather that drops from a head-dress, an insect generally found only outside but which uncharacteristically penetrates indoors, a cock that suddenly crows at midday. All these are likely to be read as alerting signs from the dead if they take place in a social context heightened by alarm or emergency.

But the most reliable signs in authenticating narratives are taken to be those of life, death, sickness and appearance themselves. A death, or persisting illness, is likely to be taken to authenticate – at least temporarily – one from among a number of narratives. Equally, the birth of children, particularly male children, is a robust counter to accusations that a man is a witch, or has otherwise betrayed his clan, or that relations with his maternal kin are awry (Figure 8.2). It is this that explains in part the otherwise odd Wahgi practice of naming a son 'Kumye': 'witch' or 'traitor'. It is to confute an accusation of treachery or witchcraft and throw it back in the face of the accuser – it is proof in an uncertain world of moral probity.

The facts of life and death, sickness and appearance are not only taken as passive indicators of the truth or otherwise but are sometimes presented as actively deployed to test narratives. Let me give an example. During my fieldwork, I was most closely associated with the two clans Kekanem and Anzkanem. Up until the early 1950s both had comprised sections of a single, larger clan. The event that precipitated their splitting had been the killing in a brawl by Anzkanem men of a Kekanem man named Simbil. This killing still preoccupied Kekanem men while I was in the field. They were concerned not simply with which of them had betrayed Simbil, thus allowing Anzkanem to kill him, but with two further points. First, they were unsure from which subclan within Anzkanem the killers had come. Secondly, they were divided as to whether the Kekanem subclan which had started the fight with Anzkanem had adequately compensated the Kekanem subclan to which Simbil himself had belonged.

For a decade following Simbil's death, Kekanem and Anzkanem remained taboo to each other, but finally they made peace, their status as two separate clans being marked by Anzkanem sending a bride to Kekanem to bear a son in Simbil's place. Intermarriage between Kekanem and Anzkanem tentatively commenced. In this situation, the facts of fertility and appearance were scrutinized in an attempt to determine answers to the outstanding questions. One Kekanem man declared that Kekanem brides had been sent in marriage to each of the Anzkanem subclans suspected of killing Simbil, and in particular to the major suspect subclan, Palakanem. 'If Palakanem were innocent', my informant continued, 'then the skin of the bride we sent them would look fine. She would have borne healthy sons and daughters'. But, as it was, her only son had died, she had but two daughters, and her own appearance was shrivelled. These facts confirmed for him that Simbil's killers lay within Palakanem subclan. Equally, the variations in fecundity and barrenness amongst the brides sent by Anzkanem to Kekanem were used to read off whether sufficient intra-Kekanem compensation had been paid.

Of course, the facts of fertility, of life, death and appearance are not stable as authenticators: childless people produce infants, glossy-skinned and healthy people later become sick. But their

Figure 8.2. *Moru and daughter, 1980. In an uncertain world, the Wahgi sometimes take the birth of children as authenticating or disproving the many conflicting narratives about past events and their capacity to influence the present.* © *Michael O'Hanlon.*

reversal, far from being seen as disproving the previous interpretation, can equally be taken as testament to the efficacy of the measures taken. Nor, equally, do births, deaths and appearance simply bring closure by authenticating uncertain narratives; rather, they are themselves also facts to be explained, around which multiple narratives will develop. But as Strathern and Stewart (2000: 147) note 'the final "backstop" of any system is likely to be the human body itself'.

Conclusion

My jumping off point at the start of this paper was with Levi-Strauss's speculations as to why certain durable artefacts – *churinga* in Aboriginal society, archives in our own – are endowed with a special or sacred character. Levi-Strauss's answer had to do with the role of these in testifying to pure historicity, to the passage of time. The first, and main, point that I have sought to make for the Wahgi is that their problematic, at least at one level, is a very different one. The Wahgi, in a sense, have too much history, not too little. As I have sought to bring out, the Wahgi world is pervaded by rival narratives, each of which purports to explain misfortunes as an enchained series whose origins lie back in relations with clanspeople, and with 'source' people like maternal kin. The Wahgi problem, then, is not to be reassured of history's existence but the very different one of sorting out the true from the false from among the range of potential histories in which they see themselves embroiled. In this situation, what is needed is an authenticator, and what I have shown is that the Wahgi turn not to durable and unchanging objects like *churinga* or archives but to the facts of life and death, to illness and appearance, to fertility and to reproduction.

My second main point is one of the concomitants of this. If we are interested in local understandings of reproduction and fertility in Melanesia, we cannot necessarily segregate them as a topic to be investigated in solitary splendour. The Wahgi, at least, view patterns of fertility and reproduction as part of a much broader universe of signs, all of which can be kaleidoscopically assembled to make sense of each other, and to determine what must have gone on in the past to affect the present in the way that it is.

Thirdly, I have sought to provide some permutations in Melanesian ethnography. The Wahgi case material provides us with an example of a rich interpretive cosmology among one of the central Highlands peoples whose lives have tended to be documented more through their exchange activities, warfare etc.

Finally, let me return to issues of history. Most of this chapter has pivoted around one kind of history: the plethora of Wahgi narratives which explain as interlinked wholes otherwise unrelated fortunes and misfortunes. But I also touched on the intersection of this history with, for example, Western biomedical history in the form of the 1940s dysentery epidemic. I want to end by asking – if not fully answering – the question of how the Wahgi explanatory system which I have outlined copes, for example, with the brute fact of population increase. The population of Komblo tribe, with whom I work, was 562 in 1952, 581 in 1955, 605 in 1957, 663 in 1962, 708 in 1965, 753 in 1967/8, 860 in 1980 and 1,200 in 1990, the last point for which I have census figures. How does this objective doubling of population over 50 years feed back into a system in which fertility and reproduction are taken to authenticate moral probity? Briefly, there are a number of partial answers. In part, the scales are disparate: individual reproductive disasters, and the drive to find an explanation for them in the terms outlined, may nevertheless co-exist with a doubling of population at the macro level. In part, as time passes, there remain a smaller and smaller number of people who recall vividly the period of pre-contact warfare, and the 1940s epidemic, to which are traced many of the originating acts. And to some degree, slight so far, the adoption of Christianity encourages a degree of recalibration, in which deaths and misfortunes are starting to be traced to unChristian anger with others and to failure to confess as well as to the activities of traitors or offended maternal kin.

Acknowledgements

I am most grateful to Tony Crook and to Dan Jorgensen with whom I have discussed aspects of this paper, and especially to Chris Ballard and Stanley Ulijaszek for comments on it. The fieldwork on which it is based was carried out with my wife Linda Frankland who, as always, has also contributed greatly to the analysis.

Notes

1. Since this paper discusses how the Wahgi read off from patterns of fertility the imprint of the past, I should acknowledge this chapter's own remote origins which lie in a seminar series on material culture and the objectification of time and history organised by the late Alfred Gell at the London School of Economics.
2. See O'Hanlon and Frankland (2003) for a fuller account.
3. This case, and further permutations of it, are discussed in O'Hanlon (1989: 64ff). As the issue remains a charged and sensitive one, the names of individuals and their clans and subclans have all been changed.

References

Barth, F. 1975. *Ritual and Knowledge among the Baktaman of New Guinea*. New Haven: Yale University Press.
———. 1987. *Cosmologies in the Making: a Generative Approach to Cultural Variation in Inner New Guinea*. Cambridge: Cambridge University Press.
Bercovitch, E. 1989. 'Mortal Insights: Victim and Witch in the Nalumin Imagination'. In *The Religious Imagination in New Guinea*, eds G. Herdt and M. Stephen, 122–159. New Brunswick: Rutgers University Press.
———. 1998. 'Dis-embodiment and Concealment among the Atbalmin of Papua New Guinea'. In *Bodies and Persons: Comparative Perspectives from Africa and Melanesia*, eds M. Lambek and A. Strathern, 210–231. Cambridge: Cambridge University Press.
Brunton, R. 1980a. 'Misconstrued Order in Melanesian Religion', *Man* 15(1): 112–128.

———. 1980b. 'Order or Disorder in Melanesian Religions?', *Man* 15(4): 734–735.

Burton, J. 1983. 'A Dysentery Epidemic in New Guinea and its Mortality', *Journal of Pacific History* 18(4): 236–261.

Connolly, B. and R. Anderson, 1987. *First Contact: New Guinea's Highlanders Encounter the Outside World*. New York: Viking Penguin.

Errington, F.K. 1988. 'Review of F. Barth, Cosmologies in the Making', *Man* 23: 766.

Gell, A. 1975. *Metamorphosis of the Cassowaries: Umeda Society, Language and Culture*. London: Athlone Press.

———. 1980. 'Order or Disorder in Melanesian Religions?' *Man* 15(4): 735–737.

Harrison, S. 1993. *The Mask of War: Violence, Ritual and the Self in Melanesia*. Manchester: Manchester University Press.

Juillerat, B. 1980. 'Order or Disorder in Melanesian Religions?' *Man* 15(4): 732–734.

———. 1992. *Shooting the Sun: Ritual and Meaning in West Sepik*. Washington: Smithsonian Institution Press.

Kelly, R.C. 1977. *Etoro Social Structure: a Study in Structural Contradiction*. Ann Arbor: University of Michigan Press.

Levi-Strauss, C. 1966. *The Savage Mind*. London: Weidenfeld and Nicolson.

Meigs, A. 1984. *Food, Sex, and Pollution*. New Brunswick: Rutgers University Press.

Muke, J.D. 1993. 'The Wahgi *Opo Kumbo*: an Account of Warfare in the Central Highlands of New Guinea', Ph.D. Thesis, University of Cambridge.

O'Hanlon, M.D.P. 1989. *Reading the Skin: Adornment, Display and Society among the Wahgi*. London: British Museum Publications.

———. 1993. *Paradise: Portraying the New Guinea Highlands*. Bathurst: Crawford House Press.

——— and L.H.E. Frankland, 2003. 'Co-present Landscapes: Routes and Rootedness as Sources of Identity in Highland New Guinea'. In *Landscape, Memory and History: Anthropological Perspectives*, eds P.J. Stewart and A. Strathern, 166–188. London: Pluto Press.

Read, K.E. 1959. 'Leadership and Consensus in a New Guinea Society', *American Anthropologist* 61: 425–436.

Strathern, A.J. and P.J. Stewart, 2000. *Arrow Talk: Transaction, Transition, and Contradiction in New Guinea Highlands History*. Kent, Ohio: Kent State University Press.

Wagner, R. 1967. *The Curse of Souw: Principles of Daribi Clan Definition and Alliance*. Chicago: Chicago University Press.

Whitehouse, H. 2000. *Arguments and Icons: Divergent Modes of Religiosity*. Oxford: Oxford University Press.

CHAPTER 9

VARIATIONS ON A THEME: FERTILITY,
SEXUALITY AND MASCULINITY IN
HIGHLAND NEW GUINEA

Pascale Bonnemère

From male initiations to pig festivals, from large-scale ceremonial exchanges to tranvestite rites, from spirit cults to fertility rituals, a great variety of ritual forms was once found on the island of New Guinea. Various rituals have tended to be studied separately by social anthropologists – with the exception of initiations and bachelor cults (see Modjeska n.d.), which are both regarded as *male* rituals, that is, organized by, performed by and intended mainly for *men*. Overall, scholars have made little attempt to look for similarities in the underlying sociology of ritual performances or for common patterns of ritual action. Consequently, spirit cults in general, and male initiations are often considered in the anthropological literature as contrasted kinds of rituals associated with different types of social organization.

As one variety of spirit cult, bachelor cults, were found only in the Western Highlands and involved a female spiritual entity with whom those young bachelors having decided to participate and who hold 'paraphernalia and secret lore as a corporate group separate from the married men of the community' (Modjeska nd: 2) entered into contact. They did so in order to enhance their growth and attractiveness, to protect themselves from their future wives and to restore overall fertility, that of the land and that of human beings, for which a number of sacrifices were also made to the ancestors.[1] Some of these groups – primarily the Enga – were famous for their large-scale ceremonial exchanges organized by prominent figures known as big-men.

By contrast, collective male initiations were compulsory for all boys of the same age-class (Godelier 1986: 183), who 'were inducted into the social group of adult men, by those men as the cult's corporate owners' (Modjeska nd: 3). These rituals focused on the novices' physical maturation and were performed among populations inhabiting the eastern parts of the Highlands of Papua New Guinea (PNG) which were characterized by antagonistic relations between men and women as well as by the absence of politically prominent male figures and of inter-group ceremonial exchanges. Among their objectives were 'to confirm men's superiority over women in the process of procreating children, and above all in the ability to produce sons ...; to establish and deepen the solidarity among ... the men belonging to one or two clans, apparently never those of the tribe as a whole' (Godelier 1986: 183). And rather than sacrifices to ancestors, 'gifts of game [were] presented by adult men to the young initiates, and with the handing of food to the onlookers, without competition, in an atmosphere of giving' (Godelier 1986: 184).[2]

Although male initiations have been regarded by some scholars as fertility cults (Whitehead 1986), I prefer to reserve the word 'fertility' for practices intended to promote 'general fertility'. For, among the Angans,[3] general fertility, of the earth, of plants and animals, is neither a concern nor a goal of ceremonial performances. The question of whether this is also true of other groups that organize male initiations, like those in the Sepik area, on the Papuan Plateau or in the Mountain Ok region, remains open. But for the Angans, male rituals are explicitly intended to bring young boys to maturity and make them into men (in a physical as well as a moral sense). Another good topic of discussion would be whether these rites could be seen instead as a preparation for marriage (which they are, *de facto*, even if this is not explicitly stated by informants), and therefore as a ritual means for assuring the reproduction of human beings. On the whole, however, there is not the slightest indication, in Anga male initiation, that might lead us to think that it is also aimed at enhancing the fertility of pigs and gardens or promoting women's fecundity (see also Lemonnier this volume). Moreover, the timing of male initiation never corresponds to a period of drought or to any other type of natural disaster that people might be attempting to cope with. People organize male initiations whenever a few dozen boys are about to reach the required age for it, and they then plant huge gardens of taro at high altitudes far away from villages to provide the necessary food associated with all ritual life in general. As Lory writes, 'Here we thus have a complementary relation between ritual and agriculture, with the latter serving as a calendar and temporal marker' (1982: 257). But again, there is no question here of promoting any sort of fertility.

In New Guinea, fertility cults proper come in either the form of Spirit cults in the Western Highlands, or 'ritualized heterosexual' practices, in a number of South Coast cultures. *Amb Kor*, the Hagen Female Spirit cult, was a circulating ritual held to renew the fertility of the cosmos. The spiritual entity made herself known to a group leader in a dream or vision; the leader would then organize the ritual ceremony of propitiation to her. If carried out properly, the Female Spirit 'would place part of her fertility powers in the cult stones that the ritual performers [had] gathered before she herself moved on to another community' (Stewart and Strathern 1998: 61). These stones were buried in the clan soil after the Female Spirit cult had been completed.

Among South Coast societies, fertility was enhanced through extra-marital heterosexual intercourse, either by wife-lending or ceremonial sexual relations between married men and prepubescent girls, or through frequent coitus in order to obtain mingled sexual fluids (Knauft 1993: 52; Lemonnier 1993: 47–48). Here, ritualized heterosexuality rather than symbolically enacted copulation was 'strongly associated with cosmological and social fertility as well as with feasting and warfare' (Knauft 1993: 53). A great deal of diversity thus exists in ritual forms designed to achieve general fertility.

In this chapter, I will focus on two major sets of rituals: bachelor cults and male initiations. I will build my argument on papers by Modjeska (nd) and Clark (1999) as well as on several recent contributions to a volume on the participation of women in New Guinea male rituals (Bonnemère, 2004). I would like to show that, although they appear to be fairly contrasted in their organization and content, these two sets of rituals can be viewed as variations on a single universal theme concerned with female reproductive capacities. My aim here is, first, to bring out this common theme underlying the different ritual performances and, second, to try to explain these patterns of variation, primarily by relating them to other aspects of social life, making a kind of 'sociology of male initiations and of bachelor cults', as Modjeska put it almost 15 years ago (nd).

Female symbolism was of course recognized long ago by scholars studying the ritual life of New Guinea men, but, as far as I am aware, it has not been the object of any comparative work across the usual dividing lines between bachelor cults, fertility cults and initiation rituals. Yet, this is the only way to reveal this common structure and to understand what form fertility – or fertility-associated themes – takes in each ritual situation.

Ankave male initiations

The Ankave ritual cycle involves three phases: two are collective and are generally organized separately (at an interval of several months), but may also be performed one after the other, a fact that corroborates the idea that they are one and the same rite. Although each phase is given a different name, which designates the main ritual act to which the boys are subjected, perforation of the septum for the first phase and rubbing with red-pandanus seeds for the second, these ceremonies, organized for a group of from 20 to 30 male children between the ages of 8 and 12 years, form a whole and are clearly different from the third and final rite. The last ceremony revolves around one young man whose first child has just been born, and who has spent the entire period of his wife's pregnancy preparing for the event by respecting dietary taboos, being relatively inactive, wearing a barkcape on his head and therefore looking different from other men. The people involved are the specialists in rituals, the young fathers who serve as surrogate brothers for the novice and those men who wish to accompany them to the forest location where it takes place.

In explanation of the two first stages of this ritual cycle, the Ankave say that it is essentially to allow boys to grow up, since they are not as lucky as girls, who mature without any outside help, and to make them strong and brave. Consequently, we would tend to label these initiations 'rites of passage', during which children pass on to another phase of their development or to adulthood. The third and last stage celebrates the procreative power of a young man since taboos are observed during his wife's pregnancy[4] and a secret rite is held in the forest between men when the baby is born. A number of objects are manipulated or planted during these different phases of the ritual cycle. These include red earth, cassowary daggers and red cordylines, which are linked to the blood and bones of a male ancestor killed in a primordial era. By this token, sacred power of the ancestor is considered to circulate among the novices' bodies and minds.

The analysis of the collective phases of Ankave male initiations is based on the rich material gathered by Pierre Lemonnier between 1987 and 2002 from Ankave male informants as well as on his eye-witness accounts of the first- and second-stage initiations in 1994, and on my own observations among the women of this group at the same time. The several third-stage ceremonies that were held during this period are not taken into account because they raise questions that go beyond the scope of the present chapter.

In short, during the few weeks of men's ritual activity, a series of actions, behaviours and ways of living can be observed, which include action on the body (perforation of the septum, blows, anointment, stretching, heating, cooling); physical ordeals (climbing a big tree, crossing a stream on a slippery log, going down narrow corridors lined with branches, running as fast as possible); psychological ordeals (sleep deprivation, death threats, lies, revelations of secrets); living in a group (sharing meals and nights in a small shelter, hunting in parties, fetching wood and water); instruction (on sharing, mutual aid, the code of ethics, taboos, proper behaviour for adult men, including not crying, self-control, self-restraint, and overcoming one's fear); scenes to be watched (theatrical miming of primeval scenes, ritualized fights between men); and imposed behaviour and attitudes (crouching in line, staying still and silent, keeping the head bowed).

During the whole time the initiates' septums are healing (the first stage of the ritual), the boys' mothers are secluded in a large collective shelter and are required to respect numerous daily behavioural constraints and food taboos, which, for the most part, are similar to those imposed on the initiates in the bush. Mothers and initiates alike must also soak their new bark capes in stream water every day at dawn before putting them on their shoulders. In the forest, next to the initiation hut, men often comment on these restrictions and particular behaviours of the novices' mothers, which they consider to be an absolute condition for the success of the male ritual. Men add that this is so 'because they gave birth to the boys in the first place'.

Many of the taboos the novices' mothers must respect are the same as those that they had to obey when pregnant. Red-pandanus juice is a food that has to be eaten by a pregnant woman, since it adds to the amount of blood she has in her body on which the fetus feeds. In effect, for the Ankave the mother is solely responsible for the intra-uterine growth of her child. At the time of initiation, however, it is the boys, and not their mothers, who consume the red-pandanus juice between the first and the second stages of the ceremonies. Moreover, when the boys eat this plant-substitute for blood, they do it in utmost secrecy[5] (Bonnemère 1996: 25–27).

The main rite of the second stage of the initiations, and the one that gives it its name, is the violent rubbing of the boys with red-pandanus seeds. The initiates are put into a tiny shelter where they are placed beside an intense fire. This shelter is built next to the entrance of a corridor of branches, both ends of which are decorated with red leaves and beaten bark, which is dyed with red-pandanus juice. Pushed by their sponsors, who are preferably real or classificatory mother's brothers, the young boys advance into the corridor, while being beaten by men posted on the outside. As they emerge, a man flings cooked red-pandanus seeds onto their face and shoulders, which he rubs in together with a reddish ochre.

I have suggested elsewhere (1996: 349) that these two adjoining frameworks of foliage are metaphors for the uterus and the vagina, respectively. The difficult progression of the initiates through the narrow passage – which, because they are pushed, can literally be called an expulsion – can be interpreted as their re-birth to a new state. The red elements at the entrance and exit of the corridor are metaphors for the blood which they believe fills the uterus and of which a small amount spills out at delivery. And like a newborn, whose head is the first body part to emerge at delivery, only the head and shoulders of the initiates are rubbed with red pandanus seeds.

When this rubbing has been completed, the boys go back to the village and are treated like newborn babies: two women cover their bodies in the yellow mud with which every infant is rubbed soon after birth. The charm employed is exactly the same on both occasions. Then the initiates distribute the rats, birds and small marsupials they have caught in the forest to their sisters and mothers, which parallels the gift of game by the husband and male kin of a woman soon after she gives birth. In both cases, these rats are called by the same name ('marsupials for a birth').

Thus I propose that initiation is a symbolic rebirth, preceded by the enactment of a gestation process (the secret consumption of red-pandanus juice). Since growth is the main concern and goal of these rituals and since, for the Ankave, the agent of the fetal growth is blood, which is provided by the mother, I suggest that gestation has been 'chosen' as the most appropriate metaphor for 'growing' boys.

It is obviously nothing new to consider male initiations as modeled on beliefs about the process whereby a new human being is reproduced, but what is amazing in this particular case is the cluster of facts which make this reference redundant. That such ritual brings about the separation between mothers and their sons is similarly quite a common idea, especially if one thinks of studies influenced by developmental psychology (Whiting *et al.* 1958: 361). Among the Ankave, this transformation of the mother–son relationship is made material by a gift of game presented to their mothers by the novices upon their return to the village. So, by the end of the ritual, the former symbiotic relationship between a mother and her son has become an exchange relationship (Strathern 1988: 222). The boy has acquired a form of autonomy, underscored by the possibility of his being an agent (in this case, of initiating the gift), and no longer simply the product of the actions of others, chief among whom is his mother. Thus in the Ankave initiation rituals, the mother–son relationship is definitely a major focus of attention (Bonnemère, in press). But if the boys can change and become somebody else, this is also because they have received some of the sacred power of the primordial male ancestor which, long ago, infused the earth, the cordylines, the bones and all these elements that now make up the sacred object which is used by the ritual expert (see Lemonnier this volume). As the main bearer of this ancestral greatness, such an object makes sacred wherever it happens to be, and influences all those present.

The Enga bachelor cult

Things are different in the Enga bachelor cult, where it is the marital relationship the young man will soon undertake that is emphasized, rather than his former connection to his mother. However, the context men cite to justify holding such rituals is similar to that of the Anga initiation described above: village life is a source of danger because there are women living in the village; sexual relations take place there; the boy's eyes necessarily see genital organs, sexual fluids, and his ears necessarily hear talk of these defiling realities. This ritual emphasizes the need for young men to cleanse their eyes, for it is through them that pollution from these areas enters and debilitates a person (Biersack 2004). And so that young bachelors may have sexual relations with their spouse and reproduce without risk, they must undergo ritual seclusion (for some, on several occasions) in the upper forest where there are no polluting influences, and where they enter into contact with a female spirit.

In Ipili (Paiela-Enga) origin myths, the *omatisia* woman (from the word for the bog-iris) is the youngest of two sisters, who decided to become a spirit woman when her elder sister married, announcing that men would seek her out when they wanted to have children. She underwent this transformation when she broke her ankle, and her blood formed the lake in which young men place bamboo tubes during the bachelor cult ritual. Wherever her hair or her legs touched the ground, there sprang up plants of the sacred flower – the bog-iris – ritually grown by men.

In both cases, the woman who had sacrificed herself to become a spirit was a virgin and, instead of bearing children, she gave birth to rituals for protecting men from women (Strathern 1970: 578; Clark 1999: 13). Concretely, she is also the source of material objects manipulated by men in the rituals and which sometimes serve as vehicles for their relationship with her.

Young Ipili men are given a set of bamboo tubes the second time they come to the cult area (Biersack 2004). They spend a great deal of time first of all preparing these, checking them, and then covering the mouth of the tubes with the bark of a special tree (a delicate operation involving manipulation and charms). The bamboo tubes are then set in swampy ground and water allowed to seep in; the bachelor's surveillance consists of interpreting the nature of this water from one ritual season to the next. The efficacy of the spirit woman's action on the boy is judged by what he sees in and at the surface of the shortest tube, the residence of this spirit. The water should be clear, in the sense that only then would it enable the bachelors to grow; if the water is limpid, it indicates that the *omatisia* woman is not menstruating. In this case the water is poured into a new tube provided by the ritual specialist. The long bamboos, on the other hand, are associated with the 'health and scale of the boys' skin' (Biersack 2004), a gauge of their growth. The young bachelors also tend bog-iris plants, and utter charms to encourage certain parts of their body (hair, eyes, shoulders) to become big and strong (Biersack 2004).

How has this spirit, with whom young bachelors have contact during their stays in that forest space not contaminated by human actions, been interpreted by anthropologists? Everybody agrees that the Spirit Woman is regarded as a wife for young bachelors, a wife who assures the growth and beauty of bachelors and protects them in their future encounters with their real wives. So, rather than focusing on the mother–son bond that has to be transformed from a symbiotic to a dual bond for the boys to be able to grow and marry in the future, the Enga bachelor cult prepares the youths for marriage by linking them to a spiritual virtual spouse whose intervention will protect them from pollution by their future real wives.

Comparative comments

In an already mentioned paper, Modjeska gives a detailed description – although based on informants' accounts and not on observation – of what happens to voluntary bachelors in the Duna *palena nane* cult, which can easily be compared with the Paiela-Enga bachelor cult.

In Duna ritual performances, bachelors also have to wash their face in the water on cold forest mornings. Here a senior bachelor called The Father, rather than each young man individually, can tell by looking at the plants who had broken a food taboo or thought about or talked to a woman (Modjeska nd: 5). The first *palena* plant was brought back from a black lake by two women. The younger woman married on the advice of the elder, who planted the *palena*. She told men that, "if they stayed with the palena, they couldn't stay with their wives, mothers or sisters" (Modjeska nd: 7).

Bamboo tubes are part of the ritual scene, as they are in more eastern locations, but, in the Duna case, their content is drunk by the bachelors. Once emptied, the tubes are placed in soft ground, two facing the rising sun and two the setting sun. 'Later, we looked (the informant says) to see if the moss was still green. If it was dry, then our hair wouldn't grow. Another time we looked again: if our tubes had refilled themselves, then our hair would grow quickly' (Modjeska nd: 5–6). Although the ritual acts and their purposes – to grow or to grow hair – are rather similar, there is no reference in Duna *palena nane* to the Spirit Woman menstruating.[6] Nevertheless, the woman who planted the *palena* and did not marry was 'the spirit wife of the bachelors – and a jealous wife at that' (Modjeska nd: 8). Modjeska writes that her relation to married men is one of hostility: 'if one happens to encounter her, he would get lost in the bush or kill himself'. Therefore, all the ingredients are present here for including the Duna *palena nane* among bachelor cults.

In the same paper, Modjeska thinks of possible links between the Kaluli male ritual and bachelor cults, because the neighboring Huli cult is where the Highlands bachelor cults reach their southwestern limits (Modjeska nd: 3). Kaluli bachelors have to be virgins and they go into the forest; at the close of the *bau a*, the senior bachelor receives the 'visit' of a spirit bride in the form of an oddly-shaped stone sent by the *memul* forest spirits. 'Not only is the *bau a* sociologically a bachelor cult [see also Clark 1999: 15], but its particular theme of the spirit bride is explicable in terms of the content and logic of the Highlands cults', Modjeska writes (nd: 4).

In short, spiritual marriage does not seem to be limited to the bachelor cults found in the Highlands but appears to be a more general symbolic reference that may occur also in so-called Highland fringe societies.

The Enga male cults of the past

The ethno-historical work that Polly Wiessner has undertaken among the Enga is interesting for a comparative study of initiations and spirit cults, because as a long-term analysis, it may help to understand the transformations that rituals have undergone over time, and consequently to shed light on the social processes at the heart of these changes.[7]

According to P. Wiessner, on the eve of first contact with Europeans, the Enga *kepele* cult was an impressive event, convened when poor environmental conditions prevailed, indicating that the ancestors were discontent. Its main purpose then was the restoration of the earth's fertility. It essentially consisted of two core rites: the building of a cult house and the initiation of young boys into the secrets of the spirit world during the *Mote* initiation rites (Tumu and Wiessner 1998: 194).

The rituals lasted five days and started with a large pig kill, which provided men with specific pieces of pork to be steamed in pits at various sites. In a small hut located inside the ritual enclosure, an expert filled a gourd with the 'water of life'. This fluid was made primarily of sugarcane juice and pork fat and, the boys were told, it was put in the gourd by the spirit woman, who sat nursing a baby in the small hut. Meanwhile, outside the sacred area, people assembled for dancing, feasting and singing.

Boys between the ages of approximately 8 and 14 years were then brought by their fathers to special *Mote* shelters built inside the sacred enclosure, where they would spend the next three days. Their diet was limited to the juice of sugarcane and the few sweet potatoes that their mothers had given them. At some point, they were given the 'water of life' to drink. As P. Wiessner explains, 'the names used ... to describe the stages of filling the gourd and drinking from it depict the symbolic weaning from the mother. The gourd-filling process was called "the pregnant woman is expecting"; the completion of this process was called "the baby is born"; and the drinking of the sacred liquid by the boys was called "the mother's breast milk is drying up"' (Wiessner 2004). As for the spell recited while the boys drank, it mentioned blood, menstrual blood and semen. By drinking this fluid given to men by the spirit woman, boys were cleansed of the effects of breast milk and close contact with women, factors that were nurturing in early childhood, but lead to stunted physical growth later in boyhood.

Inside the ritual house, the boys sat in total darkness, while outside, experts whistled to signal the presence of ghosts. After some point, two men dressed as beautiful sky women entered the cult house and revealed sacred bark paintings by the light of their torches.

After the *Mote*, some ritual experts unearthed sacred stones in shapes reminiscent of male and female genital organs and brought them into the cult house together with a basketwork figure fashioned to look like a man (*yupini*). There, copulation was simulated between the *yupini* (or the male ancestral stone) and the female stones, while spells to promote fertility were recited. After copulation, the *yupini* was fed with pork fat, and the sacred stones were greased with it. Following the rites, which were repeated over the following days, some of the cult houses were destroyed; it was believed that the goodwill of the ancestors had then been evoked and fertility would prevail. P. Wiessner puts it well:

> *Mote* rites separated boys from their mothers around the time when they would have moved from the women's to the men's house. Once given the water of life and introduced to the secrets of the spirit world, they could enter a second stage of growth to become productive members of society who participated in communal ritual for the reproduction of the tribe or clan. (2004: 168)

This short description of the *Mote* initiation as it was performed in the Enga area shows a ritual pattern which, beyond differences between specific ritual acts, appears quite reminiscent of what takes place in the Ankave initiation. In both cases, a symbolic pregnancy and birth is enacted, although by different means: the consumption of red-pandanus juice by the novices and their progression through a corridor of branches, among the Ankave, and the different phases of the filling of the gourd, among the Enga. In both cases, the mother–son bond is the focus of attention and is transformed through reference to a real birth, and, for the Enga, to breastfeeding as well.

The *Mote* was continued until sometime around the 1960s, when contact with missionaries led to the discontinuation of the entire *Kepele* cult. The importance of the *Mote* seems to have gradually waned with a ritual shift away[8] from the theme of separation from the mother to one of constructing appropriate paths to marriage and the complementarity of male–female relationships that were so essential to men's success. It was in this context that a Female Spirit was called on, as seductress and bride in the Enga bachelor cult, locally called the Sangai, to assist men in quite a different role.

Concluding remarks

The rituals examined here do indeed refer to the making of human beings, but in a variety of ways which seem to depend on the purpose of the ritual and on the identity of the male participants. When the aim is above all 'to grow' all the young boys of a local group (Ankave-Anga), it is their bodies that receive the generative substances metaphorically associated with maleness and femaleness (Bonnemère 1996: 347–348). There is no dance or sacrifice since there are neither ancestors to propitiate nor fertility to restore. In short, as far as fertility is concerned, this is a limited form connected with human beings – and maybe even with males only – and not with broader cosmological fertility. Among the Ankave, it is most clearly present in the individual-oriented phase of the ritual cycle for which it is said that what is celebrated is the young father-to-be's procreative ability. The absence of any rite on the occasion of girls' first menstruation would also limit the fertility aspect of all of the lifecycle rituals to this third phase.

Initiations also emphasize the joining of procreative substances in the body, in other words, conception and the 'making' of a human being. In fertility cults, on the other hand, men re-enact a symbolic coitus by placing objects representing the female and male genital organs so that they touch.[9] Such emphasis on sexuality rather than on conception fits with the interpretation of the spirit woman as a wife.[10] Afterwards, the sacred stones that served as a vehicle for sacrifice to the ancestors are buried in the ground for the purpose of restoring general fertility.

It may also be that these contrasting modes of action vary with forms of 'agency'. When men have recourse to a spiritual agent to attain their goals, whether it is to prepare bachelors for marriage or to restore fertility, novices do not appear in the rite re-enacting a copulation scene. It is the ritual specialist who manipulates the objects symbolizing maleness and femaleness, and uses them to act out sexual intercourse. In initiations, the boys are the direct beneficiaries of the foodstuffs (ginger, salt, taro, and so on) assimilated to surrogates for procreative substances; likewise the primary human agent[11] connected with the transformation they undergo in the course of the ritual is their real mother and not a spiritual being.

The Duna case is exemplary of a possible transitional moment: it was the senior bachelor visited by the spirit wife who received the stone-spirit bride rather than all of the bachelors, as in the Paiela-Enga bachelor cult.

In short, the present analysis points to an opposition between, on the one hand, male initiations in which the symbolism of human growth is largely predominant and in which the novices' mothers, whether they are present in person or in the form of surrogates, are essential figures of the ritual process, and, on the other hand, spirit cults in which fertility is ensured by the symbolic re-enactment of a coitus scene and in which the principal figure is a spiritual being, sometimes embodied in a stone, who occupies the position of virtual wife.

Although this is admittedly an awkward way to put it, it is as if men 'invented' a female spirit because it was unthinkable that their real wives could take part in male cults. Or to put it another way, might not the creation of this figure in men's minds be regarded as one of the manifestations of negative representations of the woman-as-wife? Ambiguity about the spirit woman's status is indicative of this tension: as Clark writes, 'closed vaginas and bachelor cult blood refer to women as sisters and as a source of wealth and fertility, but sisters cannot be "real" wives and marriage and exchange are necessary for society to form and prosper' (1999: 16). As mothers, women are admitted to the ritual process, something attested by Anga initiations; but this is not the case when the symbols being manipulated refer to sexuality rather than to the growth of human beings.

So, women could be split, so to speak:[12] as wives, they would not be allowed to appear in the male ritual context which emphasized preparation for marriage and general fertility. A spirit woman, who, as Strathern writes, is 'a symbol of gender itself, separable of physical sexuality' (1979: 46), and therefore in a way a positive wife, has therefore taken over their role. As mothers, women are regarded positively, and therefore they can figure in rituals stressing boys' growth. From this standpoint, the question of who participates, spiritual beings or flesh-and-blood women, would be connected directly with the aim of the rituals, with the male persons concerned and therefore also with their symbolic reference world (preparation for marriage, for sexuality or physical maturation). Given that, throughout New Guinea, any woman with whom a man has intercourse is a source of pollution and danger, there seem to be contrasting symbolic configurations which might explain why, in one case, real women (as mothers) are called upon and, in another, spirit women (as brides) are called upon.

We have yet to understand the conditions under which the mother–son relationship ceases to be of ritual concern and the marital relationship becomes the primary centre of interest (for the two go hand-in-hand), but P. Wiessner's ethno-historical study of the Enga (2004) may offer some suggestions, by emphasizing the change of focus that the *Mote* initiation and its encompassing *Kepele* cult have undergone over time. Thus, in the context of demographic expansion and the extension and diversification of exchange networks, might relations of affinity not take on such importance that young bachelors must be prepared ritually for the marital relationship? It is obviously hard to answer this question, but the knowledge that initiation rites used to exist in Enga society (the *Mote* rites), and that large-scale ceremonial exchanges did not occur in the groups that performed initiation rituals, would seem to point in this direction.

Modjeska echoes this, writing: 'a plausible hypothesis would be that the Duna palena complex either reflects or was functionally involved in a historical transition from restricted to generalized (bridewealth-based) exchange' (nd: 11). And finally Godelier: 'Might we then hypothesize that the development of competitive exchange and of exchange-oriented production may have led ... to the disappearance of the collective male and female initiations?' (1986: 184).

Thus the following interpretation, that a 'primal woman, often without a vaginal opening, appears to have been passed along ritual and trade pathways, undergoing various transformations in response to historical, demographic, and political economic conditions' (Clark 1999: 25) appears to be widely accepted.

In any case, it seems clear that, whether or not ritual forms in New Guinea aim to ensure general fertility – and not only the regeneration of the ground, as in the Huli *dindi gamu* ritual (Ballard 2000: 212) – the model referred to by ritual symbolism is most often that of human reproduction.[13] The differences encountered are merely variations on this theme, which I have tried to explain by looking both at the aim of the ritual and the identity of the male participants (the whole community of boys, voluntary young bachelors or married men). But beyond these variations, human reproduction in its largest sense remains the primary model that men have recourse to in the variety of ritual forms encountered in New Guinea.

Notes

1. The Kaluli – a group living on the Papuan Plateau – have been discussed by Modjeska as an interesting case for the comparative discussion of bachelor cults (nd). Schieffelin gives, as goals of the *bau*, a ritual other than those related to the young boys' body and appearance, the warding off of sickness and death and the suspension of conflicts and revenge killings among long-house communities (1982: 158–159). Schieffelin does not consider the *bau a* ritual as an initiation (1982: 194–198), but given that the boys are transformed in one way or another during their stay in the hunting lodge, it may have been both an initiation and a fertility ritual. Similarly, as mentioned in the text that follows, south-coast New Guinea societies' male initiations were linked to the complex rituals that regenerated the life-force (Lemonnier 1993: 48).
2. For a systematic and term-for-term comparison of cults and initiations, see Godelier 1986: 183–184.
3. The set of some 70,000 people known collectively as Angans is divided into 40 or 50 local groups inhabiting an approximately square-shaped territory of about 130 by 140 km that takes in parts of Morobe, Eastern Highlands and Gulf provinces. Angans speak 12 related languages and share many aspects of social organization, descent, gender relations and male rituals. The groups inhabiting the northern part of the Anga area are the most familiar due to the work of Herdt, Godelier and Mimica in the 1970s; more recently, anthropological fieldwork has been undertaken in two Southern groups as well (Bonnemère 1996, Bamford 1997), this leading to a more diverse view of Angan ways of life.
4. It is worth noting here that what is celebrated on the occasion is a man's first-time fatherhood, not his wife's first pregnancy.
5. If I have chosen to reveal practices rather than conceal them, it is mainly because of the oldest Ankave's concern that their culture be described as precisely and exhaustively as possible in these times of great change. This being said, I would like the reader to be aware of the secrecy that surrounds some of the ritual actions analyzed here and to respect it.
6. Although the lake where *palena* plants originated is said to be black, like the bloody water where Ipili young men place their bamboo tubes.
7. Here, I rely on the work of P. Wiessner: her book, *Historical Vines*, published in collaboration with Akii Tumu in 1998, and her contribution in Bonnemère 2004.

8. Anthropologists working in the Western Highlands used to say that initiations never occurred in the societies they study, but, as Wiessner showed, things may be more complex.
9. A similar re-enactment of copulation is attested among the Huli, who also organized bachelor cults (Chris Ballard, pers. comm.).
10. In south-coast New Guinea societies as well, the emphasis is on sexuality rather than on conception (Lemonnier 1993: 48).
11. Human, because we have seen that other kinds of agents are necessary to the ritual process, like plants, ochre, bones derived from a primordial being. Among the Baruya, the Sun is also an important ritual agent (Godelier 1986: 82–83).
12. A reality quite familiar to Polynesian specialists (Tcherkézoff 1994).
13. Sometimes images of human reproduction are mixed with symbolic references to the incubation of bird eggs (Gillison 1980: 146; Rohatynskyj 2004; Strauss 1990: 37).

References

Ballard, C. 2000. 'The Fire Next Time: the Conversion of the Huli Apocalypse', *Ethnohistory* 47(1): 205–225.

Bamford, S.C. 1997. 'The Containment of Gender: Embodied Sociality among a South Angan People', Ph.D. dissertation, University of Virginia.

Biersack, A. 2004. 'The Bachelors and their Spirit Wife: Interpreting the Omatisia Ritual of Porgera and Paiela'. In *Women as Unseen Characters: Male Ritual in Papua New Guinea*, ed. P. Bonnemère, 98–119. Philadelphia: The University of Pennsylvania Press (Social Anthropology in Oceania Series).

Bonnemère, P. 1996. *Le pandanus rouge. Corps, différence des sexes et parenté chez les Ankave-Anga (Papouasie Nouvelle-Guinée)*. Paris: CNRS Editions/Editions de la Maison des Sciences de l'Homme.

──, ed. 2004. *Women as Unseen Characters: Male Ritual in Papua New Guinea*. Philadelphia: The University of Pennsylvania Press (Social Anthropology in Oceania Series).

──, ed. In press. 'Des liens à denouer: l'implication des mères dans les initiations masculines des Ankave-Anga (PNG)'. In *La dimension sexuée de la vie sociale*, eds I. Théry and P. Bonnemère. Paris: Editions de l'EHESS (collection Enquête).

Clark, J. 1999. 'Cause and Afek: Primal Women, Bachelor Cults and the Female Spirit', *Canberra Anthropology* 22(1): 6–33.

Gillison, G. 1980. 'Images of Nature in Gimi Thought'. In *Nature, Culture and Gender*, eds C. MacCormack and M. Strathern, 143–173. Cambridge: Cambridge University Press.

Godelier, M. 1986. *The Making of Great Men. Male Domination and Power among the New Guinea Baruya*. Cambridge: Cambridge University Press.

Knauft, B. 1993. *South Coast New Guinea Cultures: History, Comparison, Dialectic*. Cambridge: Cambridge University Press.

Lemonnier, P. 1993. 'Le porc comme substitut de vie: formes de compensation et échanges en Nouvelle-Guinée', *Social Anthropology/Anthropologie sociale* 1(1): 33–55.

Lory, J-L. 1982. 'Les jardins Baruya', *Journal d'Agriculture Traditionnelle et de Botanique Appliquée* 3–4: 247–374.

Modjeska, N. nd. 'The Duna palena nane and the sociology of bachelor cults', unpublished manuscript.

Rohatynskyj, M. 2004. 'Ujawe: The Ritual Transformation of Sons and Mothers'. In *Women as Unseen Characters: Male Ritual in Papua New Guinea*, ed. P. Bonnemère. Philadelphia: The University of Pennsylvania Press (Social Anthropology in Oceania Series).

Schieffelin, E. 1982. 'The *bau a* Ceremonial Hunting Lodge: an Alternative to Initiation'. In *Rituals of Manhood. Male Initiation in Papua New Guinea*, ed. G. Herdt. Berkeley: University of California Press.

Stewart, P.J. and A. Strathern, 1998. 'Ritual Trackways and Fertility in New Guinea', *Journal of Ritual Studies* 12(1): 61–66.

Strathern, A. 1970. 'The Female and Male Spirit Cults in Hagen', *Man* 5, 4: 571–585.

———. 1979. 'Men's House, Women's House: the Efficacy of Opposition, Reversal, and Pairing in the Melpa Amb Kor Cult', *Journal of Polynesian Society* 88: 37–51.

Strathern, M. 1988. *The Gender of the Gift. Problems with Women and Problems with Society in Melanesia*. Berkeley: University of California Press.

Strauss, H. 1990. *The Mi-Culture of the Mount Hagen People, Papua New Guinea*, ed. G. Stürzenhofecker and A. Strathern. Ethnology Monographs 13.

Tcherkézoff, S. 1994. 'The Illusion of Dualism in Samoa. "Brothers-and-sisters" are not "Men-and-women"'. In *Gendered Anthropology*, ed. T. del Valle. London, New York: Routledge (EASA Series).

Tumu, A. and P. Wiessner, 1998. *Historical Vines*. Washington D.C., Smithsonian Institution Press.

Whitehead, H. 1986. 'The Varieties of Fertility Cultism in New Guinea: Parts I and II', *American Ethnologist* 13: 80–99; 271–289.

J. Whiting *et al.*, 1958.'The function of Male Initiation Ceremonies at Puberty'. In *Readings in Social Psychology*, eds E. Maccoby *et al.*, 359–370. New York: Holt.

Wiessner, P. 2004. 'Of Human and Spirit Women: from Mother to Seductress to Second Wife'. In *Women as Unseen Characters: Male Ritual in Papua New Guinea*, ed. P. Bonnemère. Philadelphia: The University of Pennsylvania Press (Social Anthropology in Oceania Series).

CHAPTER 10

FERTILITY AMONG THE ANGA OF PAPUA NEW GUINEA: A CONSPICUOUS ABSENCE[1]

Pierre Lemonnier

Whereas numerous New Guinea societies are (or were)[2] famous for their fertility rituals, or for the place occupied by fertility in some of their outstanding institutions and therefore in people's everyday life, the Anga of Eastern Highlands, Gulf and Morobe provinces of Papua New Guinea have lacked any such collective practice aimed at creating, maintaining or restoring general fertility. Or rather, they limited their interest in the circulation of a life force to very specific, though crucial, domains: the highly visible making of adult men and warriors during male initiations and, for some of them, the unspoken but alarming recycling of some sort of life-giving substance within clans or lineages. As for the fertility of women, pigs, gardens, trees, game or the cosmos, separately or together, the theme was strikingly absent. This situation may not be unique in New Guinea, but it contrasts strongly with the situation among neighbouring Highlander groups.

Anga groups were characterized by a tightly interlocking system of warfare, male initiations and male–female antagonism. There was no ceremonial exchange between groups, and exchanges (of women, goods, hostilities) were kept in strict balance. Anga political leaders were Great Men singled out by their hereditary functions (as ritual masters) or their skills (as great warriors, hunters or shamans) as being superior to other men, at least in the exercise of these functions (Godelier 1986: 79–99).

Male initiations were notably a time when young men's growth, physical and psychic maturation were obtained and, in this respect they were (and still are in a few Anga groups, such as the Ankave) a key institution for the reproduction of human beings.[3] The aim of such institutions was to convey supernatural powers into the novice's body: that of the Sun for the Baruya (Godelier 1986) or Sambia (Herdt 1987), that of a primordial ancestor who was at the origin of humankind, in the case of the Ankave.

Anga male initiations were clearly *not* totally independent from the 'circulation of life'. In the first place, 'unfinished' men would somehow be infertile and have no children. But, these rituals may have been also linked indirectly with the proper order of the cosmos at large. Among the Iqwaye studied by Mimica (1981), for instance, male rituals were definitely associated with the reproduction of the cosmos.[4] Limited information suggests that in the remote past among the Ankave, the ritual preparation of the dancing-ground might have involved some sort of buttressing of the earth. This part of the ritual is now discontinued, and no one recalls it anymore. In any case, this was not a central part of initiation ceremonies.

If we consider that 'fertility' is 'the quality of being fertile', the plural of which is 'productive powers', fertile meaning 'fruitful, prolific' and/or 'causing or promoting fertility' (*Shorter Oxford English Dictionary* 1973: 742) – then Sambia male initiations were linked with fertility, at least indirectly, because the Sun was equally responsible for the metamorphosis of boys ('The Sun is our father', Herdt 1987: 105), for the origin of particular important species (the red pandanus), and for the maturation of other plants (taro and yams) (Herdt 1987: 104 ff). The Baruya also said 'The Sun is our Father' (Godelier 1986: 66). However, both the Sun *and* the Moon were indirectly implicated in the fertility of the gardens, both having to be in the right place so that the gardens were neither scorched nor flooded (Godelier 1973: 362).[5]

During male rituals, however, the local emphasis was on giving 'strength' to the boys, who were seen fundamentally as warriors (and not as husbands and fathers, for instance). And, above all – as Bonnemère also argues (this volume) – Anga male initiations had nothing to do with fertility in general, nor with gardens, women, pigs or the cosmos. For this reason, I would not call these initiation ceremonies 'fertility cults'. In order to ascertain how much and how the creation, circulation and restoration of productive powers was, and continues to be, dealt with by the Anga, I shall now consider the idea of fertility among the Ankave.

Fertility among the Ankave: a fractioned concern

The three main areas in which the Ankave intervene to control some sort of fertility are the maintenance of human health, the growing of plants in the gardens and the availability of game in the forest. They express no ideas about the general circulation of something like a life-force or a fertility principle. The only doctrine (if one can call it that) related to 'fertility' concerns the animals they hunt and trap, whose capture requires the use of a magical bundle called *sambea owe*. In the case of eels, the individual charms uttered by the trapper are supposed to be heard by the masters of the eels, who ask the fish to be kind enough to enter the trap. Other kinds of hunting magic merely consist of formulae referring to some image of abundance ('May the game taken be as numerous as the Menye in their men's houses'). All these formulae are known on an individual basis. No collective action is ever undertaken for the purpose of maintaining the abundance of game. A general view is that the bush spirits *pisingen awo* may be unhappy if people hunt or trap more than they need, in which case they attack the 'spirit' (*denge*, breath, the ability to think, life-force) of the hunter, who faints or falls into a stupor.

Garden magic is also an individual matter. When opening a new garden in the forest, the gardener leaves/left one tree standing in the middle of the plot until it has/had been entirely cleared and planted. Then the tree (called *ga'wo' ika'a*, 'the tree of the bananas', referring to the main staple crop before either the sweet potato or taro *Xanthosoma* were present in the region) would be cut. The gardener pronounces a secret formula when he cuts into the bark of the tree ('The skin of the (greasy and slippery) *temongwen* fish I cut'), and another secret formula when the tree starts to fall (something like 'May you feed these plants'). The falling of the tree to the ground is thought to enhance the growth of the plants: 'The banana trees are shaken and grow well because the tree makes the ground shake when it falls.' A few charms (thought by some to enter the ground) are spoken while the first specimen of each type of plant is put in the ground. When the first plants or fruits are harvested, they must be left on the stump of the last tree felled by the gardener, or the gardener may get sick. The name of these garden products left to rot is *peaxe'*, which is the general word for the 'offering' made to express sorrow by destroying something (Speece 1987: 298), although nothing is said about this (for instance, no one knows why the gardener would get sick).

Clearly the ritual practice in this case is not aimed at creating, circulating or restoring fertility. It is aimed merely at the success of the action undertaken and, in this respect, gardening is similar to other actions on the world such as rain-making, flood-making, producing lime, or building a house. No entity is endowed with fertility: there are simply rules to be followed so that things do not fail. As is so often the case, the only comments made about these practices are something to the effect that 'such is the custom our ancestors told us to follow' (*nenge aroo' itungwain ee' teneke*, *Tumbuna bilong mipela i bin putim lo olosem* in Tok Pisin).

For the Ankave, besides the need to utter the words that must be spoken when planting a new garden, they need to worry about garden fertility because of human ill will, or the capacity to do harm. If I am jealous of some woman's garden, I will use magic to discourage her from going and working in her garden, which will then deteriorate rapidly. This type of practice is by no means restricted to gardening but is part of an array of magical practices designed to impair or curse people, things and activities (*ramexanene*, 'to hinder, impair, curse other people's action', Speece 1987: 329). Black magic of this kind is performed, for instance, so that married people will fight, or so that a sister's son's or daughter's growth will be impaired, or so that a neighbour will abandon a building site.

Maintaining people's health is another domain for which there are no positive behaviours supposed to favour the circulation of some medium associated with the idea and power of fertility. Instead one protects oneself or others against negative powers of either spirits or humans. In particular, the majority of shamans' actions are aimed at repairing mechanical damage resulting from attacks of *ombe*, invisible cannibal spirits associated with maternal parents, whose main goal is to recycle the flesh and blood of individuals back to an undifferentiated lineage pool of life-substance (from which each mother gets the blood she provides to her own children). It is also the job of the shaman to identify and diagnose attacks by human sorcerers. In the latter case, the shaman has no power to cure the sick person, for only the original sorcerer can restore his victim to health, in particular once the victim has realized that he or she may have wronged the sorcerer and decides to make compensation (but no public confession). Again, in both cases, it is not the control of fertility that is at issue, but a battle against malevolent agents that want to destroy individual lives, whether these are evil spirits or altogether ordinary human beings. In most cases, *ombe* or sorcerer attacks result from the victim's failure to respect the local rule of sharing and exchange. In other words, if someone is ill, there is a good chance that her or his moral status is responsible.

The only collective ritual concerning health among the Ankave is a protective magic aimed at keeping people's bodies safe from negative powers. Epidemics are a good time to undertake such a ritual (which, as far as I know, has no proper name). We do not know the impact on Anga populations of the dysentery epidemics that spread throughout New Guinea during the Second World War. However, influenza epidemics in the late 1960s killed almost one-tenth of the Ankave Anga population, while during the El Niño episode of 1997–1998, 77 people (around 7 percent of the population) died in few months in one Ankave valley (Lemonnier 2001). In the protective magic ritual, most of the population gathers in a given hamlet and, one by one, each individual stands between two cordylines, while a man who knows the magic formula draws marks in blue clay (*xwa'atungwen*) on the person's belly as a preventive treatment. Along with the clay and the use of pleasantly-smelling leaves, the formula (which refers to two parrots and a bird of paradise) builds a magical screen which keeps illness from entering people's bodies. Once again, this is nothing like a positive action to promote fertility, but only a defence against negative factors.

The Ankave management of what resembles productive powers can therefore be characterized in two ways. First, like so many other New Guinea societies, they link the good health of humans, gardens and game to people's individual morality. This was already formulated by Godelier (1973) 30 years ago for the Baruya: if their gardens are in bad shape, it is either because some individuals have failed to respect rules regarding their social and sexual lives, or because they have failed to cooperate with others. According to the Ankave, there is no one substance that would systematically confer fertility on people, things and the cosmos. In fact, the list of all the substances and rituals used by the Ankave for improving productive power would be a long one. What is striking here is precisely the *diversity* of the mediums at hand. For instance, a long time ago – when the ancestors of the present-day Ankave might still have comprised a Kamea clan – it was good for young warriors to eat the raw liver or 'right arm's meat' of an enemy. Nowadays, people believe a sow will carry a lot of piglets if she is fed with the *waireo'* bulb, because this bulb has stripes that look like those of piglets (Speece and Speece 1983: 126); a very thin person may be given a piece of half-cooked game, so that his or her health will improve. In another domain, the Ankave share with other Anga the idea (implicit and not commented upon) that the bodily fluids are somehow recycled when a corpse is deposited in an old garden; the development of a young woman's breasts is hastened by rubbing them with the yellowish *omexe* clay.

In this grab-bag of magic, the use of a magical bundle deserves a special place for several reasons: it is one of those cases in which the success of a future action is at stake, rather than the retrospective repair of some weakness or damage; it is also the only circumstance in which a general procedure will serve a variety of goals, i.e. various hunting and trapping techniques, as well as catching various types of game. But, above all, the magical *sambea owe* has much in common with the sacred objects looked after by the masters of the initiations.

Acting on game and boys: variations on magical bundles

At the heart of the Ankave (and, more generally, Anga) initiations lay the sacred objects used by the masters of the initiations to complete the maturation of young 8 to 13-year-old boys and to turn them into adult men, that is to say, into strong warriors able to control their fear. These objects, which the Ankave call *oxoemexe* ('man' – 'fight') or *xwoe mexe* ('great fight'),[6] are utilized, thought, experienced and even presented explicitly as the quintessence of their culture and identity. People recognize themselves in the *oxoemexe*; they consider that they themselves and their society are what they are because of the existence and use of these sacred objects. For these men, as for most Angans, such sacred objects are simultaneously what defines their own tribal identity and what defines the Anga tribes in general (those neighbouring Anga tribes which are or were locally known), in contrast to non-Anga people. The periodical 'awakening' of the *oxoemexe* places every man in a double set of relations: with all the initiates that are his contemporaries and, notably, with those boys who were his co-initiates; but also with all the deceased men whose growth, finished state and warriorhood resulted from a contact with the primordial heroes at the origin of the *oxoemexe*.

These sacred objects are also secret objects: for the women, who at most see their external envelope from afar; for the novices, who are almost unable to look around them and are stunned by the potential violence and the incredible powers locked up in these objects; and for ordinary men, who have only a vague notion of their content. Their possession defines the function of the master of the initiations: *oxo oxoemexe*, 'the man with an *oxoemexe*'; *naye tango*, 'he who got feathers'; *oxo xenej*, 'the man-mother'). In comparison with other Anga ritual experts,[7] during a ceremony, this figure's authority over the whole Ankave population is absolute, a situation which contrasts sharply with the 'every man for himself' ideology which prevails except in times of war.

The name *oxoemexe* designates both the main sacred object (a bone awl used to pierce the novices' septum), the other artifacts enclosed in the pouch – which I will deliberately not detail – as well as the tapa container itself. An *oxoemexe* is a complex object in two ways: because of the variety of things it contains, but also because of its external relations with other material elements of the ritual system it belongs to. Composite in itself, an *oxoemexe* is also associated with other material elements of the initiation rituals: other objects (a sheaf of war arrows, bamboo flutes, all sorts of clay, scented leaves), but also a series of plants which comprise the vegetal theatre where the ceremonies take place. Some of these plants are attributed a ritual efficacy comparable to that of the *oxoemexe* sacred objects.

Each of the things contained in an *oxoemexe* taken separately – for example, a particular kind of magical nut, or a cowry shell – can be also used in various non-ritual contexts. The objects associated with the *oxoemexe* also appear in various magical practices. The sheaf of war arrows gathered just before initiations was also used in war magic and can still be involved in lethal sorcery attacks today. The bamboo flutes are played ceaselessly during shamanic seances in order to provide sound-beacons for the spirit familiars travelling in the invisible part of the world in search of a stolen soul. In other words, taken individually, the objects and elements comprising an *oxoemexe* are not enough to distinguish this sacred object from other Ankave magical artifacts. It is their association and relations in the very particular context of an *oxoemexe* that are significant here.

It is crucial to note that, by its form, content and general function, the sacred pouch itself looks very much like the magical bundles used for hunting and trapping.[8] Such a *sambea owe* contains bones (an eel skull minus the lower jaw; a cassowary leg) and various animal parts (pig ungula, cassowary feathers, echidna quills, eagle claws), magical *wiamongwen* nuts (*Mucuna albertisi*) and cowry shells, which stand for the symbolic gift offered to the master of the game. When the time comes to set an eel or a marsupial trap, the magical bundle is laid opened on a platform inside the hunter's house. A magical formula is said silently while setting the trap, which evokes the cowries offered (shown) to the master of the game, or the crowded men's house of some enemy group, so that the game will be as numerous.

The bones, cassowary quills, *wiamongwen* nuts or cowries that are inside or associated with an *oxoemexe* are not sufficient to distinguish it from the artifacts used in hunting magic. In other words, it is not the visible effect of an *oxoemexe* on the novice's body which is relevant, but the meaning the Ankave attribute to that set of elements in a ritual context. What makes the specificity of that particular magical object – the *oxoemexe* sacred object – is the way the Ankave think of its action and effectiveness, the origin of the powers it contains, as well as its relations with the other elements of the rituals in which it is involved. Hence the necessity to look at the network of meanings, explicit and implicit, surrounding an *oxoemexe*. In this respect, the Ankave locate the male rituals they perform in a history (for us, a mythical history) which is largely that of the origin of the powers brought into play by the masters of the initiations and contained in the *oxoemexe* or associated with it. All of these objects are related to two remote ancestors, whose blood, bones or feathers are associated with one or several objects linked with the *oxoemexe*: an Ankave man who was at the origin of all Humanity, and a cassowary that was changed into the first real woman. The sacred *oxoemexe* therefore materializes the conjunction of two sets of powers, male and female.

The similarity between the sacred object used in male initiations and ordinary magical bundles used in hunting and trapping also reveals a shift from a sort of abundance cultivated individually in the context of hunting to the collective use of magical means aimed at completing the making of warriors. In some way then, what is good for game is good for men. It should also be noted that, in both cases, the exchange with supernatural entities is kept to a minimum. The *pisingen awo* bush spirits that act as game wardens are apparently happy with the formulae spoken so that they will free the game which is to be caught. The capture of the cassowary, this important game animal (Bonnemère 1996), results from an exchange that occurred at the very beginning of present-day humanity, when a woman and a cassowary changed places, the bird leaving the forest to become the first human woman, while the almost-human woman became a cassowary living in the forest. This exchange is a given, and there is nothing to modify or re-enact.

Similarly, there is no negotiation or any exchange with the supernatural beings whose powerful faculties are transmitted to the bodies and minds of the initiates: they are simply asked to do this and, as soon as the master of the rituals holds the right object, its power is conveyed into the boys. In other words, in order to obtain an abundance of game or the successful transformation of boys into warriors, the Anga have – for some reason – invented a set of relationships with spiritual entities in which human beings give almost nothing but some of their time and knowledge. This represents a huge difference from the way other people in New Guinea see the various relations comprising a person, or deal with the fertility and/or growth of gardens, pigs, game, initiates and women. I have written 'for some reason' because we are clearly in no position to understand this sort of cultural choice. But we can at least try to circumscribe the domain and the implications of some of these choices, to qualify and understand the unceremonial ways by which the Anga attend to fertility. To do this, I shall recall the centrality of fertility-oriented cults in two regions of New Guinea, the Highlands and the South Coast.

The Highlands: nothing like a 'pig cult', but many porcine substances related to fertility

Generalization is a dangerous game, but I risk nothing by recalling that the Highlands are famous for at least two things regarding fertility: the importance some Highlands societies gave to the organization of more-or-less regular fertility rituals and the value that all of these societies attribute to pig substances in connection with the fecundity of humans, pigs and plants.

However, the local religious systems in which the domesticated pig and various porcine substances are associated with fertility rituals vary enormously from one Highland society to another. In fact, pig fat or blood may be only one element of a ritual whose logic, meaning and function have nothing to do with an explicit representation of the domesticated pig as a medium of fertility. A well-known example is that of the *dindi gamu* (or 'earth rituals') fertility cult performed by the Huli and their neighbours, which has been described and analyzed notably by Ballard (2000) and Frankel (1986: 16–26). What the Huli were trying to do on these occasions was to redress a 'universal tendency to entropy that takes the form of earthquakes, famine, drought and flood' (Ballard 2000: 213), and impinges on the health of human beings, the size of pig herds and the quality of gardens. This decline in cosmic fertility reflects the moral order of Huli society. And the way ritual experts restore this general fertility is by 'the observance of the structures of customary knowledge (*mana*) and the structure of behavioral codes (*ilili*)' (Ballard 2000: 209). According to Frankel (1986: 20–21), pigs were sacrificed at some stage, at least on some sacred sites of the ritual. In circumstances such as epidemics, pigs were sacrificed to the ancestors or to the *dema* spirits (1986: 24). But what was to be controlled and rendered fertile was 'the flow of fluid substance' in general (Ballard 2000: 209).

Indeed, in a very general way, in the Highlands, the regeneration of the gardens, the abundance of domesticated pigs as well as the health or growth of human beings relied on various practices in which the fat or blood of pigs was used. For instance, according to Newman, the Gururumba played flutes during their pig festivals' to ensure that the vital power they control will continue to operate and produce another cycle of growth ... [they were] wrapped in a leaf representing pigs so that when [they] are played real pigs will be stimulated to copulate and reproduce themselves. From time to time, the flutes were also 'fed' by placing bits of food in them so that their power will not diminish.' The power of the flutes was in turn transmitted to wooden poles erected in a house on the ceremonial ground (Newman 1965: 68–69).

The Kuma smeared pig *blood* on the sacred digging-stick they use to prepare the ceremonial ground where the pig festival took place. This digging-stick was also used to manipulate pieces of bark that were buried in the gardens to make them fertile (Reay 1959: 152, 156–157). As O'Hanlon wrote (1992: 597), 'fat ... is a focal substance for the Wahgi, credited with almost miraculous powers of causing growth'. The Kamano and Fore regularly dipped their ceremonial flutes in pig blood. Among the Fore, pig blood could be sprinkled directly on the ground (Berndt 1962: 72; Lindenbaum 1972: 250). The Tairora too attributed healing properties to pig blood (Watson 1983: 56). As for the Mendi, and several other Western Highlands people of PNG, they fed pig blood to their ancestors, represented by stones (Lederman 1986: 180).

Pig *fat* was either eaten or smeared on living beings or things whose fertility one wanted to promote. Maring and Daribi people explicitly equated it to semen (LiPuma 1981: 275; Wagner 1967: 628). The same protective function was recognized by the Chimbu, who sprinkled the ceremonial shelter of the *bolum* spirit with pig blood and grease (Brown 1972: 49), and by the Nondugl, during the fertility rites in which the *bolim* spirit participated (Luzbetack 1954: 109). The Enga fed fat to sacred fertility stones (Tumu *et al.* 1989: 35–36). When starting a new garden, a Kuma 'expert' smeared the digging-sticks with pig grease (Reay 1959: 10). In Western Papua, young Dani novices were fed pig grease and were anointed with it, and they kept a piece of grease next to them during their ceremonial seclusion (Heider 1972: 189). Young Dani brides were rubbed with pig grease and received armbands made of pig testicles and penises to promote their health and fecundity (O'Brien 1969: 213). Warriors, cadavers, mourners and sacred stones were also among those 'things' which the Dani smeared with pig grease (Heider 1970: 107, 118, 151–152, 164, 291). They would even rub grease and pig fat on the rocks of the quarries before extracting the stones for their axes (Pétrequin, pers. comm. 1986).

I stress that the pig and porcine substances look as though they may have been a basic *medium* for conveying fertility or life-force in the Highlands. This medium was certainly not without meaning: in some way or other, consciously or not, the pig was associated with life, and often 'works' as a substitute for life. But the pig was *not* the centre of attention in fertility rituals; there was nothing like a pig cult. There were local systems of meaning which had their own logic, their own imaginary content, their own story and history. And pig blood and fat occupied a central, although not unique, place among the mediums or the substances which are manipulated to restore or promote fertility. No more, no less. Yet, it is observed that pigs were used in a great variety of ways in various fertility rituals in the New Guinea Highlands. The pig was also at the heart of a system of representations and practices that ties together peace procedures, homicide compensation and economic rivalry (Lemonnier 1990, 1991). This contrasts with the ethnographic situation on the South Coast, where enemy heads, human sexuality and sexual fluids were the basis of the fertility of the cosmos.

South-Coast New Guinea societies: or how to avoid domesticated pigs at all cost

Fertility was of great importance among the societies of South Coast New Guinea (SCNG) and depended on a symbolic complex that combined warfare and various body parts of slain enemies, the life-cycle of plants, and the fecundity of humans, none of which were mediated by domestic pigs. The basic and unremarkable idea here was that life is born from and feeds on death (Serpenti 1984: 316; van Baal 1966: 601, 753–754): death of enemy warriors, whose severed heads give life to the coconut trees; or death of an old couple, whose decreasing vital energy will ensure the rebirth of the gardens and the reproduction of human beings. This was illustrated in particular by two often co-existing practices: that of erecting 'head trees', which identify the human body with a tree, and the coconut (or the fruit of the sago palm) with a human head; and that of germinating a deceased person's coconuts on his own grave before transplanting them elsewhere (Boelaars 1981: 121, 169; Serpenti 1968: 136, 1977: 212; van Baal 1966: 601, 753–754; Williams 1936: 282, 371; Zegwaard 1959: 1039). Severed heads were notably the link between warfare and the periodic reproduction of plants. The Asmat (Zegwaard 1959: 1022), the Keraki (Williams 1936: 177) and the Marind-Anim (van Baal 1966: 676) gave the name of decapitated victims to their children as a first or second name. In various places, the taking of a head used to be a prerequisite for marriage (Boelaars 1981: 172; Landtman 1927: 248; Zegwaard 1959: 1041). No power resided in the skulls themselves (Boelaars 1981: 67; van Baal 1966: 788; Williams 1936: 284), but, with the exception of the Keraki, severed (male) heads occupied an essential place, alongside other objects or substances, in male initiations or in the rites that marked life-cycle events (Boelaars 1981: 167, 171; Landtman 1927: 161; Serpenti 1984: 314–315; Williams 1936: 281, 375; Zegwaard 1959: 1027–1028).

Another shared general idea was that heterosexual intercourse was favourable to, and even a prerequisite for, plant growth. Among the Kiwai and Marind-Anim, the maturation of all food crops depended on rituals during which these plants were placed in contact with sexual fluids gathered after human intercourse. In some of these heterosexual rituals – notably the well-known *otiv bombari* of the Marind-anim – a mixture of sexual fluids was collected from women who had to make love with several men in succession on the same night (van Baal 1966: 807 ff.; Landtman 1927: 70, 78–84, 90, 351–352).

In his book, *South Coast New Guinea Cultures*, B. Knauft designated the 'strong and complementary linkage between celebratory creation of life-force through ritual sexuality and violent taking of life-force through headhunting' (1993: 216) as a crucial aspect of fertility among these societies. His comparison has also shown that this particular emphasis on fertility was related in various ways with socio-political practices and institutions. In particular, he pointed out that, among SCNG societies, the cosmological beliefs linking the cultural creation of fertility and the circulation of the life-force with ritual sexuality and headhunting corresponded in various ways with residential patterns, political organization and forms of social inequality, and with some demographic features too: for instance, the sexually transmitted diseases resulting from the *otiv bombari* rituals depleted women's fertility, which led the Marind-anim to organize more and more of these heterosexual fertility-oriented ceremonies.

Considering the political and economic organization of SCNG societies, Knauft rightly remarked that they stood somewhere between the Big-Men and Great-Men societies. I myself have argued that some crucial aspects of these social organizations could be pinned down only if one referred to the Big Men/Great Men opposition stressed by Godelier (1986; see also Lemonnier 1995). In particular, it can be shown that when the domesticated pig is not locally seen as a possible substitute for human life, ceremonial exchanges – and social organization at large – contrast very strongly and in a meaningful way with those of Highlands Big-Men societies. This argument definitely has something to do with the local embedding of fertility in various dimensions of SCNG social life, which I will summarize here.

Since Modjeska (1982), Godelier (1986), Strathern (1982) and others (Lemonnier 1990, 1991), it is well-known that in Big-Men societies, exchanges of wealth take on the importance they have, and are the basis of 'Big-manship' because domesticated pigs can be exchanged in a whole array of social circumstances: marriage, peace ceremonies, homicide compensation, child-rearing rituals, mortuary payments, and so on. In other words, Big-Men societies are characterized by the existence of a unique sphere of exchange, in which the domesticated pig plays the role of an almost universal form of 'wealth'.

By contrast there was *no* unified sphere of exchange among SCNG societies, and it looks 'as if' they were trying very hard to distinguish between numerous types of exchanges, and 'as if' one way to achieve this partitioning of exchanges was to disregard the domesticated pig as a possible means of exchange by denying its value as a substitute for human life. In effect:

- Besides the gift of children or of a spouse that characterized peace procedures, SCNG societies compensated the life of warriors with goods (shells, dog's teeth, canoes) which differed from those (vegetal products) offered in competitive exchanges. This was the case even when peace ceremonies were organized on the same occasion as the competitive exchanges supposed to 'replace' fights (Serpenti 1968: 125). It is remarkable that the pig was totally absent from the domain of compensation.
- Large-scale ceremonial exchanges (of food) explicitly maintained peaceful relations between groups, with the idea that 'economic' competition would replace warfare, but these exchanges were separate from peace ceremonies.
- The items circulating in the competitive exchanges were vegetal products, to the exclusion of any other kind of wealth, especially pigs.

- Wealth items played only an indirect and limited role in marriage and, more generally, in relations with maternal kin and affines; there too, the pig appeared only marginally.

In SCNG societies, then, people paid compensation for life and death, using for this purpose some form of wealth of which pigs or pork were never a part. Furthermore, wealth and reference to compensation used to be equally absent from important ceremonial prestations, which were no less competitive for that. In other words, these are societies that utilize wealth to pay compensations, raise and circulate pigs, and practise competitive inter-group exchange, but *these three terms were never overtly linked in any way*.

The pig was strikingly absent from the 'wealth' used for compensation procedures. Compensation payments did exist in the South of the Island, but they took primarily the form of the exchange of children and wives; only rarely (among the Kimam and Kiwai) did they entail gifts of wealth, *none* of which ever included pigs. As far as ceremonial exchanges go, even when they were connected with peace-making (Kimam), they involved 'objects' (vegetable products) that were never used as a medium of compensation.

In sum, as far as compensation is concerned, these South New Guinea societies differed from those in other parts of New Guinea, where the gift of domestic pigs or pork was conceptualized as a gift of some kind of 'life force', or as a 'token of life', which represents a human being in marriage or death compensations. It was the wild pig which was, among others, a symbol of life, fertility and growth (but also of ardour and bravery), used in the growth rituals or the initiations. But the pig was not used as a substitute for human life, in homicide compensation or peace procedures, for instance. Furthermore, as we have seen, the pig played no role in fertility rituals in those societies.

Can it therefore be postulated that the *domesticated* pig must be attributed life-giving powers to be able to be regarded and utilized as a medium for compensating life and death? Or to put it another way, what is needed for the pig to become a precious object, and even a life-compensating item of wealth? From a technical standpoint as well as in local symbolic representations, the pig was seen, in these SCNG societies, as a semi-domesticated wild animal. The raising of pigs in captivity – in this case, of young born of domesticated females impregnated by wild males – was limited or non-existent. Williams (1936: 224) did not see any litters at all, and van Baal (1966: 406–407) states that, without exception, all domesticated pigs were wild males that had been captured. Another limiting factor were the local forms of agriculture, which was either underdeveloped or aimed largely at intergroup exchanges, for which people went to considerable additional effort (Landtman 1927: 383; Serpenti 1977: 247; van Baal 1966: 713; Williams 1936: 232–233).

Here, the emphasis fell on the identification of the warrior with a wild boar, which took many forms and was a symbol of combat and bravery (van Baal 1966: 66, 146, 149, 408; Boelaars 1981: 67, 174–175; Landtman 1927: 359; Serpenti 1977: 158). The mythology closely associated wild boars with headhunting, and more generally with warfare (Landtman 1927: 365–366; van Baal 1966: 212; Zegwaard 1959: 1021).

It is tempting then to speculate that these animals must be intensely socialized by labour (the case in Big-Men societies) and, as it so happens, by the labour of women, who not only look after the pigs but basically cultivate their gardens in order to feed them. In other words – and this is pure speculation in an attempt to understand an absolute correlation – an animal that, in its wild state, symbolizes strength and maleness, would be transformed into an item of wealth, and above all wealth recognized as a life-substitute, or 'token of life', suitable for use in compensation and exchange, only when domesticated by female labor – and particularly when it reproduces in captivity (Lemonnier 1992a).

Pinning down an absence

As we have seen, in contrast to the Highlands, the Ankave have nothing which approaches a fertility ritual. They never use porcine substances in any practice related to fertility, and human sexuality is not a model for fertility either; as P. Bonnemère (this volume) explains, it is instead motherhood that provides the main model for the growth of the boys during the male initiations in this part of New Guinea. A striking feature of the rituals related to growth, health or abundance is their *diversity*. Not only may the magical formulae people know and use differ from one individual to another, but the very principles on which the effectiveness of these rituals is based do not have much in common. This sort of partitioning is reminiscent of the separation the Ankave make in another area of social life, the use of wealth in exchange.

But the Ankave have none of these exceptional forms of currency that must be kept and not used for exchange, so that they may serve to gauge the ultimate value of those less precious valuables that do circulate (Godelier 1998, A. Weiner 1992). Alternatively, they possess secret, sacred objects that are the source of the circulation of a common substance among all men. This substance brings the adult male to completion, something women do not know how to do, thus creating, over the years, bands of warriors that have watched and continue to watch over the valley. The symbol and agent of these ties between the living, the dead and the primordial heroes is the *oxoemexe*, which, in the field of male rites, plays the double role of actor and visible agent of the crucial relations between society and the beings that inhabit the cosmos, which other Melanesian peoples entrust to shell-monies or to carvings used in rites associated with life and, more spectacularly, with death. For the analyst, the Ankave's magical tool-box is also one of those *objets séparés* which is used in this part of the world to fill a 'lack' and to finish human beings on various occasions (Breton 1999). It carries out this indispensable imaginary task by visibly manifesting to the eyes and the imaginations of everyone – or at least the members of the male community – this predisposition to fractionate which Melanesians seem to attribute to human beings (Strathern 1988), which is itself inherent in the propensity of humans and spirits to incorporate parts of other people or to detach parts of their own bodies or selves (Lemonnier 1992b).

When considering the nature of the shell moneys offered at various times in the life-cycle of a person, the Ankave situation appears to differ fundamentally from that observed in island Melanesia as well as in the Highlands of New Guinea. Most of these shell artifacts are primarily means of payment. They are indeed 'precious objects' (Panoff 1980). But none of them could be said to be an 'iconic' representation of the person (Battaglia 1983, Breton 2000), or a 'fragment of clan body' (Breton 2002: 22), no more than there is a strict equivalence between the types of meat offered to various parents over the life-cycle of the person for the marriage or initiation, of whom it is received. These gifts go mainly to the maternal kin of the bride or novice (Bonnemère 1996: 169–171), but they are a way to acknowledge the rights and obligations of the person whose status is thus celebrated, rather than some sort of representation of the person themselves.

To summarize, wealth (shells, pig meat, *Pangium edule* sauce), is not absent from the gifts by which the Ankave underscore all sorts of social relations. But, in most cases, a particular item of wealth must be given in a particular context, and only in this one. In other words, compared to Highlands societies, Great-Men societies do not lack the use of wealth; they lack the use of a general medium of exchange (Lemonnier 2002). And the reason they lack it is that the exchanges in which objects are indirectly and/or incipiently equated to human beings are compartmentalized. In turn, this may parallel their incipient use of the pig as a substitute for life. The pig is used, and in small quantities, only as part of a brideprice or as a gift to a maternal uncle upon the initiation of his nephew, but it is not suitable for homicide or war compensations. Nor, as we have seen, is it suitable as a fertility-enhancing substance either. By contrast, when the principal medium of compensation is or was *also* the object of exchanges, and when this object is a pig, which can be re-exchanged, the domesticated pig naturally acquires the general appeal and nature that it enjoys in the Highlands. And, as is indirectly (but clearly) demonstrated by the SCNG case, it ensures or ensured this key role only where it is/was locally used as a substitute for life. Furthermore, in societies where the pig is used as such a substitute and exchanged in homicide compensation, peace ceremonies and the like, it is *also* an important medium in fertility rituals.

In the Highlands, the pig appeared as the main, if not only, medium of fertility. But, at the same time – and although they are only the *means* necessary to the establishment of life-circulating procedures – pigs and gardens became included among those 'things' whose reproduction is equated with that of the society as a whole: in other words, pigs are both a means to attain fertility and one of those 'items' which must be made to flourish by means of fertility rites.

Anga people are spectacularly *unconcerned* about fertility, and one hypothesis that may reasonably be advanced to explain this conspicuous absence is that they do not develop (as in the Ankave case) or they *lack* altogether (in the Baruya case) the general equivalence between persons and objects that underlies the representation of the pig as a substitute for life. I don't know why this is. But, in the absence of an explanation, it may at least be noted, at least for the sake of further research, that the Anga also have limited their emphasis on the partibility of the person to the sphere and relations surrounding the making of boys in male rituals, and to the use of sacred objects by ritual specialists.

But the Ankave are not one of those societies where 'the objects which constitute wealth function not only as *substitutes for persons*, for human beings, but also as *substitutes for sacred objects*, which are the ultimate source of all human power' (Godelier 1998: 149). And a hypothetical reason for this is that, in this case, the partibility process concerns the dismantling of primordial heroes *and* the way substitutes of part of themselves, notably their blood, 'finishes' the boys according to a symbolic pattern that mimics human reproduction and motherhood (Bonnemère, this volume). As already mentioned, in the sphere of compensation for a human life, if there are few exchanges and these are kept strictly separate, it is because the pig is not a substitute for a human life. By the same token, the Ankave would have no 'reason' to associate gardens and women with the production of pigs, nor to regard the fertility of this trio – women, gardens and pigs – as a basis for the reproduction of their society and cosmos as a whole. But they do link the magic technique and the help they seek to get game from the forest, to the techniques used to create warriors.

Notes

1. This chapter contains some information whose disclosure could harm the 'living culture' of the Ankave, to use the formula once given to me by a high-ranking PNG official from the National Research Council. I consider it to be the reader's responsibility to use any information contained in this piece of anthropological work with the greatest care, and only within the scientific arena.
2. To avoid the constant use of both past and present, I have generally adopted the past tense instead of the sometimes misleading ethnographic present. Those few Ankave rituals described in the present tense are indeed still performed.
3. Our fieldwork among the Ankave started in 1982, with funds from the Centre National de la Recherche Scientifique and the constant assistance of the Papua New Guinea Institute of Medical Reseach (Goroka).
4. 'It is the notion that men are the source of fertility which, through cohabitation with women, is generated in the world and cosmos at large' (Mimica 1981: 56).
5. However, local garden magic makes no connection with this sort of cosmic equilibrium. Baruya garden magic is about protection against rats, spiders and other insects, not about the place of the Sun and Moon (Godelier 1973, and pers. comm. 2002).
6. The Baruya *kwaimatnie* derives its name from *kwala*, 'man', and *yitmania*, 'to lift the skin, to grow' (Godelier 1986: 81).
7. With the possible exception of the Kapau-Kamea, where most if not all of the rituals seem to have been in the hands of ordinary men. But I must confess that we lack information here. My own slim data go back to 1980.
8. This is also true of the Baruya *kwaimatnie* (D. Lloyd, pers. comm. 1996).

References

Ballard, C. 2000. 'The Fire Next Time: the Conversion of the Huli Apocalypse', *Ethnohistory* 47(1): 205–225.

Battaglia, D.1983. 'Projecting Personhood in Melanesia: the Dialectics of Artefact Symbolism on Sabarl Island', *Man (N.S.)* 18: 289–304.

Berndt, R.M. 1962. *Excess and Restraint*. Chicago: University of Chicago Press.

Boelaars, J.H.M.C. 1981. *Head-Hunters about Themselves. An Ethnographic Report from Irian Jaya, Indonesia*. The Hague: Martinus Nijhoff.

Bonnemère, P. 1996. *Le pandanus rouge. Corps, différence des sexes et parenté chez les Ankave-Anga (Papouasie Nouvelle-Guinée)*. Paris, CNRS Editions/Editions de la Maison des Sciences de l'Homme.

Breton, S. 1999. 'Le spectacle des choses. Considérations mélanésiennes sur la personne', *L'Homme* 149: 83–112.

——. 2000. 'Social Body and Icon of the Person: a Symbolic Analysis of Shell Money among the Wodani, Western Highlands of Irian Jaya', *American Ethnologist* 26(3): 558–582.

——. 2002. 'Monnaie et économie des personnes', *L'Homme* 162: 13–26.

Brown, P. 1972. *The Chimbu. A Study of Change in the New Guinea Highlands*. Cambridge (Mass.): Schenkman Publishing Company.

Frankel, S. 1986. *The Huli Response to Illness*. Cambridge: Cambridge University Press.

Godelier, M. 1977. 'The Visible and the Invisible among the Baruya of New Guinea'. In *Marxist Perspectives in Anthropology*, ed. M. Godelier, 196–203; 238–239. Cambridge: Cambridge University Press.

——. 1986. *The Making of Great Men. Male Domination and Power among the New Guinea Baruya*, transl. Rupert Sawyer. Cambridge/Paris: Cambridge University Press/Editions de la Maison des Sciences de l'Homme.

——. 1998. *The Enigma of the Gift*, transl. Nora Scott. Chicago/Cambridge: University of Chicago Press, Polity Press.

Heider, K.G. 1970. *The Dugum Dani: a Papuan Culture in the Highlands of West New Guinea*. New York, Viking Fund Publications in Anthropology, No 49.

Heider, K. 1972. 'The Grand Valley Dani Pig Feast: a Ritual of Passage and Intensification in New Guinea', *Oceania* 42(3): 169–197.

Herdt, G.H. 1987. *The Sambia. Ritual and Gender in New Guinea*. New York: Holt, Rinehart and Winston.

Knauft, B.M. 1993. *South Coast New Guinea Cultures. History, Comparison, Dialectic*. Cambridge: Cambridge University Press.

Landtman, G. 1927. *The Kiwai Papuans of British New Guinea*. London: Macmillan and Co.

Lederman, R. 1986. *What Gifts Engender. Social Relations and Politics in Mendi, Highlands Papua New Guinea*. Cambridge: Cambridge University Press.

Lemonnier, P. 1990. *Guerres et festins. Paix, échanges et compétition dans les Highlands de Nouvelle-Guinée*. Paris: Editions de la Maison des Sciences de l'Homme.
_____. 1991. 'From Big Men to Great Men. Peace, Substitution and Competition in the Highlands of New Guinea'. In *Big Men and Great Men. Personifications of Power in Melanesia*, eds M. Strathern and M. Godelier, 7–27. Cambridge: Cambridge University Press.
_____. 'Pigs as Ordinary Wealth. Technical Logic, Exchange and Leadership in New Guinea'. In *Technological Choices. Arbitrariness in Technology from the Neolithic to Modern High Tech*, ed P. Lemonnier, 126–156. London: Routledge.
_____. 1992b. 'Couper-coller: attaques corporelles et cannibalisme chez les Anga de Nouvelle-Guinée'. *Terrain* 18: 87–94.
_____. 1993. 'Le porc comme substitut de vie: formes de compensation et échanges en Nouvelle-Guinée'. *Social Anthropology* 1: 33–55.
_____. 1995. 'Fertile Chimeras', *Pacific Studies* (Book Review Forum), 18(4): 155–169.
_____. 2001. 'Drought, Famine and Epidemic among the Ankave-Anga of Gulf Province in 1997–98'. In *Food Security for Papua New Guinea*, eds R.M. Bourke, M.G. Allen and J.G. Salisbury, 164–167. Canberra: Australian Centre for International Agricultural Research.
_____. 2002. 'Women and Wealth in New Guinea'. In *People and Things. Social Mediations in Oceania*, eds M. Jeudy-Ballini and B. Juillerat, 103–121. Durham, N.C.: Carolina Academic Press.
Lindenbaum, S. 1972. 'Sorcerers, Ghosts and Polluting Women: an Analysis of Religious Beliefs and Population Control', *Ethnology* 11(3): 241–253.
LiPuma, E. 1981. 'Cosmology and Economy among the Maring of Highland New Guinea', *Oceania* 51(4): 266–285.
Luzbetack, L.J. 1954. 'The Socio-religious Significance of a New Guinea Pig Festival', *Anthropological Quarterly* 27(3): 59–80; 27(4): 102–128.
Mimica, J. Omalyce. 1981. 'An Ethnography of the Ikwaye View of Cosmos'. Doctoral dissertation, Australian National University, Canberra.
Modjeska, N. 1982. 'Production and Inequality: Perspectives from Central New Guinea'. In *Inequality in the New Guinea Highlands Societies*, ed A. Strathern, 50–108. Cambridge: Cambridge University Press.
Newman, P.L. 1965. *Knowing the Gururumba*. New York: Holt, Rinehart and Winston.
O'Brien, D. 1969. 'Marriage among the Kona Valley Dani'. In *Pigs, Pearlshells, and Women. Marriage in the New Guinea Highlands*, eds R.M. Glasse and M.J. Meggitt, 198–234. Englewood Cliffs, N.J.: Prentice-Hall.
O'Hanlon, M. 1992. 'Unstable Images and Second Skins: Artefacts, Exegesis and Assessments in the New Guinea Highlands', *Man (N.S.)* 27(3): 587–608.
Panoff, M. 1980. 'Objets précieux et moyens de paiement chez les Maenge de Nouvelle-Bretagne', *L'Homme* 20(2): 6–37.

Reay, M. 1959. *The Kuma: Freedom and Conformity in the New Guinea Highlands*. Melbourne: Melbourne University Press on behalf of the Australian National University.

Serpenti, L. 1968. 'Headhunting and Magic on Kolepom', *Tropical Man* 1: 116–139.

_____. 1977. *Cultivators in the Swamps. Social Structure and Horticulture in a New Guinea Society (Frederik-Hendrik Island, West New Guinea)*. Assen/Amsterdam: van Gorgum.

_____. 1984. 'The Ritual Meaning of Homosexuality and Pedophilia among the Kimam-Papuans of South Irian Jaya'. In *Ritualized Homosexuality in Melanesia*, ed G.H. Herdt, 292–336. Berkeley: University of California Press.

Speece, R. 1987. *Angave-English Dictionary*, typescript. Ukarumpa: Summer Institute of Linguistics.

_____ and M. Speece 1983. *Angave Anthropology Sketch*, typescript. Ukarumpa: Summer Institute of Linguistics.

Strathern, A.J. 1982. 'Witchcraft, Greed, Cannibalism, and Death. Some Related Themes from the New Guinea Highlands'. In *Death and the Regeneration of Life*, eds M. Bloch and J. Parry, 111–133. New York: Pergamon.

Strathern, M. 1988. *The Gender of the Gift. Problems with Women and Problems with Society in Melanesia*. Berkeley: University of California Press.

Tumu, A., P. Minini, A. Kyangali and P. Wiessner 1989. *A View of Enga Culture*. Madang: KPI Publishing.

van Baal, J. 1966. *Dema. Description and Analysis of Marind-Anim Culture (South New Guinea)*. The Hague: Martinus Nijhoff.

Wagner, R. 1967. *The Curse of Souw. Principles of Daribi Clan Definition and Alliance in New Guinea*. Chicago/London: University of Chicago Press.

Watson, J.B. 1983. *Tairora Culture: Contingence and Pragmatism; Anthropological Studies in the Eastern Highlands of New Guinea. Vol 5*. Seattle/London: University of Washington Press.

Weiner, A.B. 1992. *Inalienable Possessions. The Paradox of Keeping While-Giving*. Berkeley: University of California Press.

Williams, F.E. 1936. *Papuans of the Trans-Fly*. Oxford: Clarendon Press.

Zegwaard, G. 1959. 'Headhunting Practices of the Asmat of Netherlands New Guinea', *American Anthropologist* 61: 1020–1041.

INDEX

A
adoption, 8, 136–155
 and agency, 148, 149–152
 as gift, 143, 149, 153
 and inheritance, 137, 139, 140
 as 'legal fiction', 153, 154
 and nature/culture distinction, 136–137, 154
agency, 212
age-specific fertility rate, 100
agriculture, 69
ancestors, 202, 204, 207, 210–212
Anga, 9, 202, 207, 213, 215n, 218, 220–226, 234, 238
Ankave, 9, 204–207, 211, 212, 215n, 220–226, 232–234
Arawe, 56
Asmat, 229
Ataka, J., 6,7
Australia, 2, 13, 14, 54, 92, 182–183
Austronesia, 3, 173

B
bachelor cults, 201, 203, 207–210, 212, 213, 216n
Baining, 133
Baluan island, 6, 91, 93, 94, 95, 96, 104
Barth, F., 183
Baruya, 219
Bayliss-Smith, T.P., 4, 6
Bedamuni, 7, 8, 112–133
Beek, A.G. van, 123, 127, 131

Bell, J., 68
Bentley, G.R., 116
Berndt, R.M and C.H., 15
Betzig, L., 111
Bigman 202, 230
Biology, 161–164
Birth, 8, 176–178, 196
 cohorts, 44
 spacing, 115, 131
 rates, 25, 71, 98
Bismarck Archipelago, 60
Bloch, M., 132
Blurton Jones, N.G., 116, 129
body, 222, 228
Bonnemère, 9
Boone, J.L., 116, 133
Borgerhoff Mulder, M., 111
breastfeeding, 32, 70
breastmilk, 211
bush spirits, 225

C
carrying capacity, 104, 105, 107
cash-cropping 85, 86
cassowary, 224–225
census, 75
ceremonial exchanges, 202, 214
ceremony, 64
Cheyne, A., 33
child growth, 205
childlessness, 27
children, 111
 adoption of, 121, 132, 133
 mortality of, 115, 121, 130

subsidizing reproduction of
  parents, 116, 120, 121, 133
treatment of disadvantaged,
  120–121, 128
value as spouses 127
value of, 119–122, 126, 128
work of, 116–119, 120, 121
Chimbu, 227
Chinnery, E.W., 56
Christianity, 199
Churinga, 182–183
cognition, 159–65
colonialism, 2, 5, 47, 53–65, 67, 70–75
compensation, 230–231, 234
conception, 8, 173
contact history, 92
contract labour(er), 93, 99
cooking, 192
Coombe, F., 19
cordilyne, 222
crude birth rate, 98
crude death rate, 98

D
Dani, 227
Daribi, 227
death, 173, 191, 194
  rate, 98
Demian, M., 8
demography
  and change, 131
depopulation, 3, 20–24, 38
  and fertility, 24–27
developmental psychology, 207
Diamond, J., 13
diarrhoea, 70
diet, 210
disease, 14, 23, 26, 60, 64, 74, 80
  respiratory, 15
domestication, 232
Draper, P., 116
Duna, 209, 213, 214
Durrad, W.J., 4
Dwyer, P.D., 7, 112, 113, 117, 119, 121, 122, 123, 126, 127, 131, 132, 133

E
Eastern Highlands, 10
economics, 80, 81, 102
economy, 58
education, 83, 93
egalitarianism, 111, 122, 125, 129
Enga, 202, 207–214, 227
epidemics, 222
Etoro, 185
exchange, 218, 221, 225, 230–231, 234

F
family
  extended, 117
  planning, 98, 102, 103, 104, 106
  reproduction of, 126
Fajans, J., 133
fatal impacts, 4, 53, 64, 86
fecundity, 202, 228
fertility, 2, 32, 70, 72, 82, 100, 126–129, 186, 188, 196, 202–204, 210–214, 234–235
  age-specific rate, 100
  Bedamuni women, 113–116
  cults 202–204
  decline, 35–37, 40
  and income, 84
  Kubo women, 113–116
  Mediums, 222, 227, 231
  rituals, 226, 231
Fiji, 2, 13, 18, 26
Foley, R.A., 111
food, 58, 167, 213, 230
  security, 69, 74
  sharing, 121
Fore, 227

G
garden, 105, 219, 220, 222, 226, 228
gardening, contribution by children, 117
Gell, A., 183
genealogy, 27, 30–31
gestation, 206

Godelier, M., 130 202, 214, 215n, 216n
Goldberg, T., 116
gonorrhoea, 102
Gosden, C., 5, 6
Gowaty, P.A., 111
Great Men, 218, 234
growth
  physical, 211, 213
Guadalcanal, 41
Gulf Province, 10
Gururumba, 227

H
Hadza, 116
Hagen, 203
Hawkes, K., 116, 129
headhunting, 228–229
health, 79, 188, 220–222, 232
  care, 77, 131
Highlands, 210, 226–228, 233–234
Highland fringe, 9, 210
history, 225
Hogbin, H.I.P., 3–5, 23
horticulture, 85
housekeeping servant, 93, 99
Huli, 209, 214, 216n, 226
hunting, 220, 224–225, 232

I
Identity, 223
  of groups, 125, 127
  of individuals, 125, 126
  of parents, 126
income, 82
independence, 67
infant mortality rate, 93, 94
infection, 5, 39, 41
infertility, 193
influenza, 40, 60
initiation, 173–176, 201, 203–207, 210–214, 218, 219, 223–226, 231, 233, 235
intensification, 122, 126
Iqwaye, 219
Ivens, W., 19, 22

J
Jasienska, G., 116
Jenike, M.R., 111
jobless migrant, 100

K
Kaluli, 209, 215
Kamano, 227
Kamea, 222
Kelly, R.C., 112, 131
Keraki, 229
kinship, 113, 121–122
kinship, mother's brother and sister's son, 124, 132–133
Kiwai, 229
Knauft, B.M., 112, 203
Kramer, K.L., 16, 133
Kubo, 7,8, 112–133
Kuma, 227
!Kung, 116

L
labour, 60, 67, 72, 85
  recruitment, 31, 59
Lak, 8, 159–179
land, rights of access, 124, 125, 127, 132
law, 154–155
Lemonnier, P., 9, 10, 130, 202, 203, 205, 207, 215, 216
Levi-Strauss, C., 182–183, 197
life
  cycle, 164–167
  expectancy at birth, 93, 94
life-giving substance, 218, 220, 229

M
Mace, R., 116
magic, 175, 191, 220, 221–224
  bundle, 225, 233
  formula, 220–222, 224
Maher, R.F., 75
malaria, 40, 46, 70, 79
  control, 98
Matankor, 91, 94
Manus Province, 91, 94, 95, 98, 102, 104

Marind-anim, 229
Maring, 227
markets, 86, 94
marriage, 27, 111, 113, 120,
    123–126, 214
  and elopement, 124
  childless, 27–30
  delayed exchange, 125
  fabricating status as exchange sister, 131–132
  immediate exchange, 125, 130
  patterns, 187
  polygyny, 128
  and residence patterns, 124
  responses to discord, 125
Massim, 138
Maya, 116
McArthur, N., 21
measles, 14, 26, 73–74
Melanesia, 1–3, 10, 15, 17, 26, 55, 148, 159, 183, 198, 233
Mendi, 227
menstruation, 208
Micronesia, 15
middle ground, 54
migration, 6, 72, 77, 81, 90, 94, 96, 99
Minnegal, M., 7, 112, 113, 117, 119, 121, 122, 123, 126, 127, 131, 132, 133
miscarriage, 37
missionaries, 72
missions, 61, 64
mobility, 121
Modjeska, N., 201–203, 209, 210, 214, 215n
Moorehead, A., 53
morality, 222, 226
Morobe Province, 10
mortality, 24, 27, 35, 42, 44, 73, 82, 93, 111
  infant, 44
  under-five years, 98
mortuary rites, 167–173
Mosko, M., 159

N
New Britain, 5, 56–64

New Caledonia, 2
Nutrition, 67, 74
  Survey Expedition, 79

O
Objects (sacred), 223–224, 233
Oceania, 11, 13
O'Connell, J.F., 116, 129
O'Hanlon, M., 6, 9, 227
Ohtsuka, R., 6
Ontong Java, 4, 6
orphans, 120–121

P
pacific, 13, 20, 27
Paliau movement, 93
pandanus, 211
partible person, 233–234
Passismanua, 56
peace-making, 230–231
Perelik, 6, 91–106
pig
  festival, 185–187
  fat, 226
pigs, 226–228, 230
  avoidance of, 119
plantations, 58, 60, 99
polygyny, 81
Polynesia, 15
population
  change, 31
  decline, 5, 71
  density, 73, 91, 104, 105, 106
  and fertility, 69–70
  growth, 76
  increase, 75, 96–8
  pressure, 100, 104, 106
pork, 192
  fat, 211
pregnancy, 43, 80
psychological factor, 23, 35, 55, 71

Q
quarantine, 41

R
racial suicide, 55
Rarotonga, 14

religion, 55, 183
remittance, 90
reproduction
  biological, 110–111, 123–130
  cultural, 72
  social, 110–111, 123–130
reproductive history, 43
reproductive success
  and male status, 129, 133
  and hunting, 133
return migration, 96, 106, 107
ritual, 63
  male, 219
  sexuality, 229
Rivers, W.H.R., 1, 3, 5, 28, 30, 32, 55–56, 64
Roberts, S.H., 3–5

S

sago, 68–69, 79, 85, 228
  production, contribution by children, 117–118
Schneider, D.M., 154
Schouten, W.C., 92
Second World War, 7, 41, 75, 93
Sellen, D.W., 116
settlement, 22
Seventh-Day Adventist, 94
sexually transmitted disease, 3, 6, 24, 36–38, 102
Shaw, R.D., 112, 131
Shih, C., 111
siblinghood, 144–145
Simbo, 27–30
Skinner, G.W., 129
social complexity, 113, 122, 126, 129
social relations, production of, 122
socio-biology, 111, 121, 130, 133
Somerville, B., 20
Sørum, A., 123, 127
sorcery, 189
source people, 187, 190
South Coast New Guinea, 3, 4
Sperber, D., 132
spiritual female entity, 201, 203, 207–213

Standen, V., 111
stillbirth, 44
Strathern, A., 133, 207, 208, 213
Strathern, M., 130, 147–148
strength, 188
Suau (Milne Bay Province), 8, 137–153
subsistence, 68, 85, 91, 104–105, 111, 117–119
  work, 131
survivorship, 67, 74

T

teknonyms, 126
temporal orientation, 126, 127
Torres Strait Islands, 28
total fertility rate, 6, 7, 67, 81, 83–4, 87, 101, 102
trade, 34, 59, 72, 79, 86, 92
tuberculosis, 79
Turke, P., 111

U

undernutrition, 70
urbanism, 76, 84, 90, 94, 99, 103
uterus, 206
Ulijaszek, S.J., 5, 6
D'Urville, D., 15, 16

V

vagina, 206, 213–214
Vella Lavella, 27–30

W

Waghi, 9, 183–199
wantok, 90, 91, 105, 106
warfare, 62, 69, 72, 198
wealth, 148, 213, 230–231, 233
welfare, child, 144, 150–151
Western Highlands, 9
Wiessner, P., 210, 211, 214, 216n
Williams, F.E., 68
witchcraft, 188–189
Woodyard, V., 112
Wopkaimin, 73
  work, 116
World Health Organization, 46